Behind the Enclosure

Behind the Enclosure

Life and Care of Kangaroos in Captivity

Rafeal Mechlore

Spectra Enterprise

CONTENTS

INDEX

INTRODUCTION

Kangaroos, with their famous bouncing strides and particular marsupial pockets, are loved images of the Australian outback. While these striking animals are inseparable from the huge and untamed scenes of their local land, a critical number wind up living behind the nook walls of zoological foundations, safe-havens, and natural life parks all over the planet. "Behind the Nook: Life and Care of Kangaroos in Imprisonment" dives into the complex universe of hostage kangaroos, investigating the multifaceted equilibrium of care, preservation, and public commitment that shapes their lives inside restricted spaces.

1. **Prologue to Hostage Kangaroo Populaces**
1. **The Confounding Kangaroo**
   Kangaroos, having a place with the family Macropodidae, are marsupials remarkably adjusted to the Australian landmass. Containing a few animal categories, including the Red Kangaroo (Macropus rufus), Eastern Dim Kangaroo (Macropus giganteus), and Western Dim Kangaroo (Macropus fuliginosus), these herbivorous marsupials have developed to flourish in different environments, from parched deserts to rich forests. Known for their strong rear legs, long tails, and particular jumping velocity, kangaroos are biologically fundamental as well as significant of Australia's normal legacy.
2. **The Call of Imprisonment**

The choice to house kangaroos in imprisonment emerges from a mind boggling exchange of variables. While a people might end up in safe-havens because of wounds, stranding, or recovery needs, others are important for overseen rearing projects pointed toward saving hereditary variety and adding to species preservation. Zoos and natural life parks, in their job as instructive establishments, give a stage to public commitment, encouraging mindfulness and comprehension of these remarkable marsupials.

II. Life Inside the Walled in area: Care and Cultivation

1. **The Significance of Moral Cultivation**
   Really focusing on kangaroos in imprisonment requires a nuanced comprehension of their organic and conduct needs. Moral cultivation rehearses are foremost to guarantee the physical and mental prosperity of these creatures. From nook plan and dietary contemplations to veterinary consideration and advancement exercises, each feature of hostage kangaroo the executives is fastidiously wanted to impersonate regular circumstances as intently as could really be expected.
2. **Nourishment and Dietary Contemplations**
   In imprisonment, giving a healthfully adjusted diet that reflects the kangaroo's normal scrounging propensities turns into a key test. Specialists team up to foster eating regimens wealthy in stringy vegetation, imitating the differed plant species saw as in nature. The meaning of dietary variety stretches out past sustenance, affecting stomach wellbeing, dental prosperity, and by and large physiological equilibrium.
3. **Veterinary Consideration and Wellbeing The executives**

Keeping up with the wellbeing of hostage kangaroos requires a proactive way to deal with veterinary consideration. Normal check-ups, preventive measures against infections, and quick reactions to clinical issues are necessary parts of wellbeing the executives programs. The intricacies of treating marsupials, from conceptive wellbeing to pocket related concerns, require particular veterinary skill.

III. Protection Through Imprisonment: Reproducing Projects and Hereditary Variety

1. **The Job of Hostage Rearing in Preservation**
   Hostage kangaroo populaces contribute fundamentally to worldwide preservation endeavors. Overseen rearing projects, frequently planned among zoological foundations and natural life associations, expect to save hereditary variety and forestall the deficiency of significant characteristics inside species. Examples of overcoming adversity of hostage rearing prompting the delivery and support of wild populaces highlight the basic job these projects play in species recuperation.
2. **Challenges in Hostage Kangaroo Rearing**

Notwithstanding the victories, hostage kangaroo rearing projects face moves novel to the species. Factors like regenerative physiology, social elements, and hereditary similarity impact reproducing results. Tending to these difficulties requires persistent exploration, coordinated effort, and a profound comprehension of the complexities of kangaroo regenerative science.

IV. Normal Medical problems in Hostage Kangaroos

1. **Perceiving and Tending to Wellbeing Difficulties**
   Hostage conditions present explicit wellbeing challenges for kangaroos, going from outer muscle issues emerging from restricted space to pressure related conditions. Veterinary groups and guardians work pair to recognize and address wellbeing concerns instantly. Techniques for sickness counteraction, the executives of wounds, and mental prosperity are essential parts of comprehensive medical care.
2. **Preventive Medical services Measures**

Preventive medical care estimates assume a vital part in guaranteeing the drawn out prosperity of hostage kangaroos. Inoculation programs, normal wellbeing evaluations, and proactive administration of potential stressors add to sickness anticipation and generally speaking versatility. The convergence of preventive medical care and moral contemplations directs the advancement of conventions that focus on the government assistance of individual creatures.

V. The Human Component: Job of Overseers and Veterinarians

1. **Overseers as Watchmen**
   The human component in the existence of hostage kangaroos is significant. Overseers, frequently shaping profound bonds with the creatures under their charge, become gatekeepers of their prosperity. The obligations incorporate actual consideration as well as improvement exercises, conduct perceptions, and encouraging social cooperations to guarantee a top notch of life.
2. **The Aptitude of Veterinarians**

Veterinarians work in marsupial medication carry an abundance of mastery to hostage kangaroo care. Their job stretches out past conventional veterinary consideration to include research, conduct examination, and the advancement of creative medical services conventions. Joint efforts among guardians and veterinarians make a complete structure for the balanced administration of hostage kangaroo populaces.

VI. Conduct Perceptions and Exploration

1. **The Complexities of Kangaroo Conduct**
   Concentrating on the way of behaving of hostage kangaroos offers important bits of knowledge into their mental capacities, social designs, and reactions to natural upgrades. Social perceptions, frequently worked with by progressions in innovation, add to refining farming works on, upgrading enhancement programs, and tending to likely stressors.
2. **Concentrating on Normal Conduct in Bondage**

Endeavors to imitate normal ways of behaving inside the limits of imprisonment are integral to advancing the psychological and profound prosperity of kangaroos. Perceptions of brushing designs, social associations, and regenerative ways of behaving guide the advancement of nooks and the executives rehearses that line up with the species' intrinsic impulses.

VII. Research Commitments to Kangaroo Preservation

1. **Propelling Information for Preservation**
   Research tries inside hostage settings stretch out past quick attention to add to more extensive preservation objectives. Concentrates on regenerative physiology, hereditary variety, and sickness the board in hostage populaces illuminate techniques for species preservation in nature. The collaboration between logical exploration and functional applications highlights the interconnectedness of hostage and wild protection endeavors.
2. **Moral Contemplations in Conduct Studies**

As social examinations advance, moral contemplations become fundamental. Adjusting the journey for information with the government assistance of individual creatures includes taking on painless examination strategies, focusing on encouraging feedback, and regarding the independence of the subjects. Moral rules guarantee that examination adds to preservation without compromising the prosperity of the review subjects.

VIII. Instructive Effort and Public Mindfulness

1. **A Stage for Protection Schooling**
   Hostage kangaroo populaces act as ministers for their wild partners, connecting with the general population in a remarkable instructive encounter. Zoos and untamed life parks influence the charming allure of kangaroos to pass on messages about preservation, territory security, and the interconnectedness of biological systems. Instructive effort turns into an incredible asset for ingraining a preservation ethic among guests.
2. **Job of Hostage Kangaroos in Training**

The hostage climate offers a controlled setting for instructive projects that feature the natural significance of kangaroos. Intuitive shows, directed visits, and interpretive presentations give potential open doors to guests to interface with these marsupials and gain a more profound comprehension of the difficulties they face in nature. The job of hostage kangaroos in schooling stretches out past the nook to move a feeling of stewardship for biodiversity.

IX. Connecting with General society in Preservation Endeavors

1. **Past Diversion: Encouraging Preservation Awareness**
   While hostage kangaroos without a doubt engage guests, their job reaches out past simple exhibition. Zoological establishments mean to encourage a more profound association by effectively including people in general in protection endeavors. Intelligent projects, volunteer open doors, and resident science drives engage guests to add to preservation, changing aloof onlookers into dynamic supporters.
2. **Adjusting Diversion and Schooling**

The fragile harmony among diversion and schooling represents a test for hostage offices. Striking the right harmony guarantees that guests are enraptured by the regular ways of behaving of kangaroos while at the same time engrossing preservation messages. Moral contemplations in show plan, interpretive signage, and directed collaborations assume a critical part in accomplishing this sensitive harmony.

X. Protection Endeavors and Future Headings

1. **Worldwide Preservation Objectives**
   The preservation of kangaroos, whether in bondage or the wild, is unpredictably connected to worldwide objectives. Environmental change, territory obliteration, and human-natural life clashes rise above public lines, requesting cooperative arrangements. Worldwide participation, directed by moral contemplations and a promise to biodiversity protection, shapes the direction of future preservation endeavors.
2. **Preservation Difficulties and Open doors**
   Challenges loom not too far off, from the heightening effects of environmental change to the fragile harmony between human requirements and natural life protection. Exploring these difficulties requires a comprehensive methodology that coordinates logical development, local area commitment, and moral contemplations. Amazing open doors emerge as innovative headways, economical land the executives rehearses, and the aggregate will to defend the eventual fate of kangaroo populaces.
3. **Protection Morals and Social Obligation**
   Moral contemplations structure the bedrock of dependable preservation rehearses. From the treatment of individual creatures in imprisonment to the fuse of Native information and values, moral rules guarantee that protection endeavors line up with standards of social and natural obligation. Finding some kind of harmony between protection objectives and the prosperity of both human and untamed life populaces turns into a demonstration of the moral underpinnings of the preservation development.
4. **Protection Training and Promotion**
   Instructing the general population about the difficulties confronting kangaroos

and other untamed life is essential for cultivating a protection disapproved of society. Past instruction, support turns into an impetus for change. Enabling people to become advocates for kangaroo preservation includes furnishing them with the information, apparatuses, and stages to impact strategies, support protection drives, and partake in aggregate endeavors.

5. **Protection Endeavors and Future Headings**

As we stand at the intersection of preservation endeavors, the future heading is molded by an aggregate obligation to the prosperity of kangaroos and the biological systems they possess. From embracing creative exploration techniques and reasonable practices to supporting for strategies that focus on biodiversity, the street ahead requires joint effort, flexibility, and a common vision for an agreeable concurrence among people and kangaroos.

1. **Brief overview of kangaroos in the wild**

   Kangaroos, the quintessential images of Australia, are marsupials that have adjusted particularly to the tremendous and different scenes of the landmass. Known for their strong rear legs, unmistakable bouncing step, and notable pockets, kangaroos address an interesting part of Australian biodiversity. This outline digs into the normal history, conduct, and natural meaning of kangaroos in their wild territories.

   ****1. Species Variety:**

   Australia is home to an assortment of kangaroo animal types, each adjusted to explicit conditions across the landmass. The three biggest and most notable species include:

   Red Kangaroo (Macropus rufus): The biggest marsupial, perceived for its ruddy earthy colored fur, the red kangaroo occupies dry and semi-bone-dry districts.

   Eastern Dark Kangaroo (Macropus giganteus): Found in the eastern and southern pieces of Australia, this species lean towards timberlands, forests, and green regions.

   Western Dark Kangaroo (Macropus fuliginosus): Disseminated across south-western Australia, the western dim kangaroo flourishes in seaside locales, shrub-lands, and forests.

   Moreover, other more modest kangaroo species, like wallabies and tree kangaroos, possess explicit specialties inside the different Australian biological systems.

   ****2. Territory and Appropriation:**

   Kangaroos show a noteworthy capacity to adjust to a scope of territories, from bone-dry deserts to waterfront forests. Their appropriation traverses the whole landmass, with various species inclining toward explicit locales. Red kangaroos, for example, rule the bone-dry inland regions, while eastern dim kangaroos flourish in additional mild environments along the eastern seaboard. This flexibility has permitted kangaroos to lay out a presence in essentially every kind of Australian biological system.

3. Novel Conceptive Variations:

One of the most unmistakable highlights of kangaroos is their conceptive methodology. As marsupials, kangaroos bring forth generally lacking youthful, which then proceed with their improvement in the mother's pocket. Female kangaroos have the striking skill to postpone the improvement of an incipient organism until ecological circumstances are ideal. This physiological variation guarantees that the joey is naturally introduced to a climate helpful for its endurance.

4. Social Construction and Conduct:

Kangaroos are by and large friendly creatures, showing complex social designs and ways of behaving. They structure bunches known as crowds or troops, drove by a prevailing male. The social elements inside these gatherings shift among species. Red kangaroos, for example, frequently structure free relationship, while eastern dark kangaroos might display more firm friendly designs.

The notable bouncing stride, known as pentapedal velocity, is a central quality of kangaroo development. The strong rear legs push them forward in a progression of jumps, covering huge distances productively. This method of headway isn't just energy-productive yet in addition permits kangaroos to really explore their different living spaces.

5. Dietary Propensities:

Kangaroos are herbivores with a particular stomach related framework appropriate for handling sinewy plant material. Their eating regimen principally comprises of grasses, forbs, and bushes. The capacity to remove greatest sustenance from bad quality vegetation is pivotal for endurance in the frequently supplement unfortunate conditions they possess. Kangaroos have front teeth for trimming vegetation, while their complex stomachs help in the maturation and processing of extreme plant strands.

6. Biological Job:

Kangaroos assume a crucial natural part in molding Australian environments. As herbivores, they impact vegetation elements through touching, which can affect plant organization and design. Furthermore, their brushing propensities add to diminishing the gamble of fierce blazes by limiting the amassing of dry vegetation. Kangaroos, as prey species, likewise support the environmental equilibrium by giving a food source to hunters, for example, dingoes and birds.

7. Difficulties and Dangers:

While kangaroos have adjusted to a scope of conditions, they face different difficulties in nature. Environment misfortune because of urbanization and agribusiness, rivalry for assets with domesticated animals, and human-untamed life struggle are huge dangers. Furthermore, environmental change presents gambles

by modifying vegetation designs, water accessibility, and fueling outrageous climate occasions.
8. Preservation Status:
As of current appraisals, kangaroo species are by and large not delegated jeopardized. In any case, confined dangers, populace declines, and worries about reasonable administration rehearses have prompted continuous preservation endeavors. Observing and understanding the effect of human exercises on kangaroo populaces are fundamental for guaranteeing their proceeded with endurance in nature.

2. **Importance of understanding and improving kangaroo care in captivity**

Kangaroos, with their notable jumping walks and particular marsupial attributes, spellbind the hearts of individuals all over the planet. While these symbolic animals flourish in the tremendous scenes of Australia, a critical number end up in imprisonment, shared with the consideration of zoological establishments, safe-havens, and untamed life parks. Understanding and further developing kangaroo care in bondage isn't just an obligation; it is a urgent endeavor with expansive ramifications for the government assistance of these marsupials, preservation endeavors, and public mindfulness. This investigation dives into the complex elements of kangaroo care, looking at the significance of moral cultivation, the difficulties looked in hostage conditions, the job of veterinary consideration, and the more extensive ramifications for preservation and training.

1. **Moral Farming: Guaranteeing the Prosperity of Hostage Kangaroos**
1. **Impersonating Indigenous habitats**
   Moral kangaroo care starts with establishing conditions that impersonate the normal environments of these marsupials as intently as could really be expected. Fenced in areas ought to give adequate room to bouncing, brushing, and social communications. Vegetation, haven, and substrate decisions add to an invigorating and enhancing climate that supports regular ways of behaving.
2. **Dietary Contemplations**
   Understanding the wholesome requirements of kangaroos is central to their prosperity. Hostage diets ought to reproduce the variety of vegetation viewed as in the wild, advancing actual wellbeing and forestalling diet-related issues. Coordinated efforts between nutritionists, overseers, and veterinarians are fundamental to create adjusted and species-explicit eating regimens.
3. **Social Improvement**
   Kangaroos are astute and social creatures that flourish with mental and actual feeling. Conduct enhancement exercises, like searching difficulties, climbing designs, and social collaborations, add to the general government assistance of

hostage people. Improvement forestalls weariness as well as supports the declaration of normal ways of behaving.

4. **Regenerative Contemplations**

Understanding the regenerative science of kangaroos is basic for overseeing hostage populaces. Female kangaroos have remarkable conceptive transformations, including undeveloped diapause, which permits them to defer the improvement of an incipient organism until great circumstances win. Regenerative projects ought to line up with the normal reproducing patterns of these marsupials, taking into account factors like social elements and hereditary variety.

II. Challenges in Hostage Conditions

1. **Wellbeing Difficulties**
   Hostage conditions present explicit wellbeing challenges for kangaroos. Outer muscle issues, stress-related conditions, and helplessness to illnesses require proactive veterinary consideration. Understanding the effects of restricted space on actual wellbeing and carrying out preventive measures are basic to tending to these difficulties.
2. **Social Elements**
   Kangaroos are social creatures that structure complex social designs in nature. Hostage conditions might introduce difficulties in reproducing regular social elements. Cautious perception and the board of gatherings, taking into account factors like age, sex, and individual characters, add to solid social communications.
3. **Mental Prosperity**
   The mental prosperity of hostage kangaroos is unpredictably connected to their capacity to communicate normal ways of behaving. Weariness, dissatisfaction, and stress can emerge in restricted spaces. Giving open doors to investigation, critical thinking, and social commitment are pivotal for advancing mental prosperity.
4. **Moral Contemplations in Hostage Rearing**

Hostage rearing projects are fundamental for hereditary variety and species protection. Notwithstanding, moral contemplations emerge, including the expected pressure of bondage, the effect on regular ways of behaving, and the obligation to forestall overbreeding.

Finding some kind of harmony between rearing for protection and guaranteeing the government assistance of individual creatures requires cautious moral contemplations.

III. Veterinary Consideration and Wellbeing The board

1. **Particular Veterinary Ability**
   Giving ideal medical services to kangaroos in imprisonment requests particular veterinary skill. Marsupial medication envelops exceptional contemplations, from regenerative physiology to dental wellbeing. Cooperative endeavors among overseers and veterinarians guarantee extensive wellbeing the executives programs that address the particular necessities of kangaroos.
2. **Preventive Medical care Measures**
   Preventive medical care estimates assume a critical part in keeping up with the strength of hostage kangaroos. Customary wellbeing appraisals, immunization programs, and proactive administration of potential stressors add to illness avoidance and in general flexibility. Moral rules guide the advancement of medical services conventions that focus on the government assistance of individual creatures.
3. **Tending to Normal Medical problems**

Hostage conditions might open kangaroos to explicit medical problems, for example, pocket related concerns, foot issues, and dental issues. An exhaustive comprehension of the normal medical problems permits overseers and veterinarians to execute preventive measures and quick intercessions when required.

IV. Propagation and Reproducing Projects

1. **Preservation Importance**
   Hostage kangaroo reproducing programs hold critical preservation esteem. By protecting hereditary variety and forestalling the deficiency of significant attributes, these projects add to the drawn out endurance of kangaroo species. Examples of overcoming adversity of hostage rearing prompting the delivery and support of wild populaces highlight the basic job these projects play in species recuperation.
2. **Challenges in Hostage Kangaroo Reproducing**
   In spite of the triumphs, hostage kangaroo reproducing programs face provokes exceptional to the species. Factors like regenerative physiology, social elements, and hereditary similarity impact rearing results. Tending to these difficulties requires consistent examination, coordinated effort, and a profound comprehension of the complexities of kangaroo conceptive science.
3. **Examples of overcoming adversity and Leap forwards**

Examples of overcoming adversity in hostage kangaroo reproducing grandstand the positive effect of cooperative endeavors. From the introduction of solid joeys to the fruitful renewed introduction of hostage reared people into the wild, these accounts feature leap forwards in conceptive science, cultivation practices, and protection procedures.

V. Social Perceptions and Exploration

1. **Propelling Information Through Social Examinations**
   Concentrating on the way of behaving of hostage kangaroos offers significant bits of knowledge into their mental capacities, social designs, and reactions to ecological upgrades. Conduct perceptions, frequently worked with by progressions in innovation, add to refining farming works on, upgrading improvement programs, and tending to possible stressors.
2. **Concentrating on Regular Conduct in Imprisonment**
   Endeavors to imitate normal ways of behaving inside the bounds of bondage are key to advancing the psychological and profound prosperity of kangaroos. Perceptions of brushing designs, social connections, and conceptive ways of behaving guide the improvement of walled in areas and the executives rehearses that line up with the species' intrinsic impulses.
3. **Research Commitments to Kangaroo Preservation**
   Research attempts inside hostage settings stretch out past quick attention to add to more extensive protection objectives. Concentrates on regenerative physiology, hereditary variety, and sickness the executives in hostage populaces illuminate techniques for species protection in nature. The collaboration between logical exploration and functional applications highlights the interconnectedness of hostage and wild protection endeavors.
4. **Moral Contemplations in Conduct Studies**

As social examinations advance, moral contemplations become principal. Adjusting the journey for information with the government assistance of individual creatures includes taking on harmless examination strategies, focusing on uplifting feedback, and regarding the independence of the subjects. Moral rules guarantee that examination adds to preservation without compromising the prosperity of the review subjects.

VI. Instructive Effort and Public Mindfulness

1. **A Stage for Preservation Instruction**
   Hostage kangaroo populaces act as representatives for their wild partners, connecting with people in general in a novel instructive encounter. Zoos and natural life parks influence the magnetic allure of kangaroos to pass on messages about preservation, territory security, and the interconnectedness of biological systems. Instructive effort turns into an amazing asset for imparting a preservation ethic among guests.
2. **Job of Hostage Kangaroos in Schooling**

The hostage climate offers a controlled setting for instructive projects that feature the environmental significance of kangaroos. Intelligent shows, directed visits, and interpretive presentations give open doors to guests to interface with these marsupials and gain a more profound comprehension of the difficulties they face in nature. The job of hostage kangaroos in training reaches out past the nook to move a feeling of stewardship for biodiversity.

VII. Drawing in General society in Preservation Endeavors

1. **Past Amusement: Encouraging Preservation Awareness**
   While hostage kangaroos without a doubt engage guests, their job stretches out past simple exhibition. Zoological organizations mean to encourage a more profound association by effectively including people in general in protection endeavors. Intuitive projects, volunteer open doors, and resident science drives enable guests to add to protection, changing aloof observers into dynamic backers.
2. **Adjusting Amusement and Training**

The sensitive harmony among diversion and schooling represents a test for hostage offices. Striking the right harmony guarantees that guests are enthralled by the regular ways of behaving of kangaroos while at the same time engrossing preservation messages. Moral contemplations in show plan, interpretive signage, and directed connections assume a vital part in accomplishing this fragile harmony.

VIII. Preservation Endeavors and Future Bearings

1. **Worldwide Preservation Objectives**
   The preservation of kangaroos, whether in bondage or the wild, is unpredictably connected to worldwide goals. Environmental change, living space annihilation, and human-untamed life clashes rise above public lines, requesting cooperative arrangements. Worldwide collaboration, directed by moral contemplations and a pledge to biodiversity protection, shapes the direction of future preservation endeavors.
2. **Protection Difficulties and Open doors**
   Challenges loom not too far off, from the heightening effects of environmental change to the sensitive harmony between human requirements and untamed life protection. Exploring these difficulties requires an all encompassing methodology that coordinates logical development, local area commitment, and moral contemplations. Amazing open doors emerge as mechanical headways, supportable land the executives rehearses, and the aggregate will to defend the fate of kangaroo populaces.
3. **Protection Morals and Social Obligation**
   Moral contemplations structure the bedrock of capable protection rehearses. From the treatment of individual creatures in imprisonment to the consolidation of Native information and values, moral rules guarantee that preservation

endeavors line up with standards of social and natural obligation. Finding some kind of harmony between protection objectives and the prosperity of both human and untamed life populaces turns into a demonstration of the moral groundworks of the preservation development.

4. **Preservation Training and Support**
   Teaching the general population about the difficulties confronting kangaroos and other natural life is critical for cultivating a protection disapproved of society. Past schooling, support turns into an impetus for change. Engaging people to become advocates for kangaroo protection includes furnishing them with the information, apparatuses, and stages to impact arrangements, support preservation drives, and partake in aggregate endeavors.
5. **Preservation Endeavors and Future Bearings**

As we stand at the junction of preservation endeavors, the future heading is molded by an aggregate obligation to the prosperity of kangaroos and the environments they possess. From embracing creative examination strategies and reasonable practices to pushing for strategies that focus on biodiversity, the street ahead requires joint effort, flexibility, and a common vision for an amicable conjunction among people and kangaroos.

C. Purpose of the book

The reason for the book, "Behind the Nook: Life and Care of Kangaroos in Imprisonment," is to give a complete investigation of the multi-layered universe of hostage kangaroo the executives. Through a nuanced assessment of moral cultivation, veterinary consideration, reproducing programs, conduct perceptions, and preservation drives, the book intends to reveal insight into the mind boggling balance expected to guarantee the prosperity of kangaroos living inside the limits of zoological establishments, safe-havens, and untamed life parks.

1. **Encouraging Comprehension and Mindfulness**
   At its center, the book tries to encourage a more profound comprehension of the difficulties and obligations related with really focusing on kangaroos in bondage. By digging into the complexities of their regular ways of behaving, physiological transformations, and social elements, the story intends to furnish perusers with bits of knowledge that rise above the limits of walled in areas. Through this getting it, the book means to develop compassion and appreciation for these notable marsupials.
2. **Moral Rules for Hostage The board**
   The moral treatment of hostage kangaroos is a focal topic all through the book. By framing the standards of moral cultivation, the story gives guardians, veterinarians, and preservationists with a guide for guaranteeing the physical and mental prosperity of kangaroos. This incorporates contemplations for nook

plan, dietary practices, conceptive administration, and conduct advancement exercises that line up with the species' regular impulses.

3. **Overcoming any barrier Among Science and Practice**
   "Behind the Nook" fills in as a scaffold between logical information and functional applications in hostage kangaroo care. By incorporating the most recent exploration discoveries, veterinary conventions, and social perceptions, the book furnishes experts and guardians with proof based approaches. This combination of science and practice is significant for refining cultivation strategies, propelling veterinary consideration, and tending to the special difficulties presented by kangaroos in imprisonment.
4. **Preservation Suggestions**
   The book highlights the fundamental job of hostage kangaroo populaces in more extensive protection endeavors. By investigating the meaning of rearing projects, hereditary variety protection, and the potential for renewed introduction into the wild, the account positions hostage kangaroos as dynamic supporters of the preservation of their species. It stresses the interconnectedness among hostage and wild populaces, supporting for an all encompassing way to deal with kangaroo preservation.
5. **Instructive Effort and Promotion**
   Instruction and backing are indispensable parts of the book's motivation. By exhibiting the magnetic allure of kangaroos in bondage, the story turns into a device for drawing in the general population in preservation drives. The book supports zoological foundations, untamed life parks, and instructors to use hostage kangaroo populaces as envoys for their wild partners. It stresses the job of public mindfulness in cultivating a feeling of obligation towards untamed life and their living spaces.
6. **Moving Future Stewardship**
   "Behind the Fenced in area" tries to rouse people in the future to become stewards of biodiversity. Through stories of fruitful rearing projects, protection forward leaps, and moral kangaroo care, the book means to impart a feeling of hopefulness and organization in perusers. It urges people to effectively partake in preservation endeavors, whether through supporting drives, chipping in, or upholding for arrangements that focus on the government assistance of kangaroos and their biological systems.
7. **Exploring Difficulties and Embracing Amazing open doors**

As kangaroo populaces face different difficulties, from territory misfortune to environmental change, the book gives a guide to exploring these impediments. It energizes a proactive and versatile way to deal with kangaroo the executives, stressing the open doors introduced by innovative headways, cooperative exploration, and reasonable protection rehearses. By tending to difficulties and jumping all over chances, the book graphs a course towards a future where kangaroos flourish both in the wild and in imprisonment.

Chapter 1

Kangaroo Species and Characteristics

Kangaroos, notable marsupials local to Australia, are a different gathering of animal types showing intriguing qualities. This special family, Macropodidae, incorporates four primary genera: Macropus, Lagorchestes, Petrogale, and Onychogalea. Inside these genera, different species show astounding variations to their current circumstance, way of life, and conceptive techniques. This broad investigation plans to dig into the assorted universe of kangaroo species, revealing insight into their scientific classification, life structures, conduct, and biological importance.

Scientific categorization and Arrangement

The family Macropodidae envelops an extensive variety of kangaroo species, each with unmistakable elements and biological specialties. The sort Macropus, the most notable, incorporates a few famous animal groups like the Red Kangaroo (Macropus rufus), Eastern Dark Kangaroo (Macropus giganteus), and Western Dim Kangaroo (Macropus fuliginosus). These species are additionally arranged in light of variables like size, hue, and natural surroundings inclinations.

Another class, Lagorchestes, addresses the more modest, more spry kangaroos. The jeopardized Grouped Rabbit Wallaby (Lagorchestes fasciatus) is an outstanding animal types inside this family. Petrogale involves rock wallabies, adjusted to rough landscapes, while Onychogalea incorporates rabbit wallabies, more modest family members with unmistakable attributes.

Life structures and Transformations

Kangaroos are famous for their special transformations, finely tuned north of millions of long stretches of development. Their most particular element is, obviously, their strong rear appendages and long, solid tails. The rear appendages are specific for bouncing, the essential method of movement for kangaroos. This energy-proficient technique empowers them to cover enormous distances with negligible exertion, a vital benefit in the parched scenes they possess.

The skeletal design of kangaroos is adjusted to help their bouncing way of life. The lengthened rear appendages, intertwined lower leg bones, areas of strength for and give security during fast jumps. The tail likewise fills in as a pivotal adjusting organ, permitting kangaroos to make sharp turns and explore differed landscapes really.

Moreover, kangaroos show sexual dimorphism, with guys commonly bigger than females. The arms of kangaroos are more modest and less created, as they are not utilized for impetus. Their front appendages are basically utilized for prepping and touching.

Kangaroos are herbivores, and their dentition is adjusted to crushing intense vegetation. They have sharp incisors for editing grass and molars intended for productive biting. The dental equation of kangaroos mirrors their herbivorous eating regimen and is a fundamental part of their environmental specialty.

Marsupial Proliferation

One of the most captivating parts of kangaroo science is their special conceptive framework. Kangaroos are marsupials, meaning they bring forth somewhat lacking live youthful, which then, at that point, keep on creating outside the belly in a pocket. The female kangaroo has two uteri and can postpone the advancement of an incipient organism until natural circumstances are ideal.

After a short development period, the small, immature posterity, known as joeys, are conceived. The mother directs the jellybean-sized joey into her pocket, where it proceeds to develop and create, connecting itself to a nipple for sustenance. This pocket based improvement is a particular marsupial characteristic, permitting kangaroos to adjust to unusual ecological circumstances.

The pocket gives a solid climate to the joey's initial turn of events, offering insurance and admittance to sustenance. As the joey develops, it steadily invests more energy outside the pocket, at last leaving it however returning for sustenance until completely weaned.

Kangaroo Conduct and Social Construction

Kangaroos display a scope of ways of behaving formed by their natural specialty and social elements. The social design of kangaroo bunches changes among species. The Red Kangaroo, for instance, frequently frames free, open gatherings known as crowds, while the Eastern Dim Kangaroo is bound to be tracked down in more modest, affectionate nuclear families.

Pecking order is a critical part of kangaroo social, not entirely settled by size and age. More established, bigger guys are commonly prevailing and have special admittance to assets and mates. Prevailing guys frequently show forceful ways of behaving, like boxing with their forelimbs, to lay out and keep up with their status inside the gathering.

Correspondence among kangaroos is fundamentally non-vocal. They utilize a scope of actual prompts, including body pose, tail position, and looks, to pass on data about their expectations and economic wellbeing. Furthermore, pounding the ground with the rear feet is a typical type of correspondence, making others aware of expected dangers.

Kangaroos are crepuscular, meaning they are generally dynamic during day break and nightfall. This conduct assists them with keeping away from the intensity of the day, a critical transformation in the frequently cruel Australian climate. During the day, kangaroos might rest in concealed regions, moderating energy for their nighttime and crepuscular exercises.

Biological Importance

Kangaroos assume a vital part in the biological systems they occupy, impacting vegetation elements and adding to supplement cycling. As herbivores, they shape plant networks through their taking care of propensities, controlling the development of specific plant species and advancing biodiversity.

Their capacity to cover enormous distances looking for food and water adds to seed dispersal, supporting the recovery of plant species. Kangaroos are very much adjusted to the parched and semi-dry districts of Australia, where their brushing and perusing exercises assist with keeping an equilibrium in vegetation.

In spite of their biological significance, kangaroos face different dangers, including natural surroundings misfortune, rivalry with animals for assets, and, now and again, separating because of seen overpopulation. Preservation endeavors are fundamental to guarantee the endurance of various kangaroo species and keep up with the environmental equilibrium they add to.

1.1Different species of kangaroos

Kangaroos, having a place with the family Macropodidae, are a captivating gathering of marsupials local to Australia. Inside this family, different species exhibit an assorted cluster of transformations, ways of behaving, and environmental specialties. This broad investigation plans to dive into the unmistakable qualities of various kangaroo species, revealing insight into their scientific categorization, morphology, conveyance, and environmental jobs.

Scientific categorization and Grouping

The family Macropodidae envelops a wide assortment of kangaroo species, each adjusted to explicit conditions and showing one of a kind elements. The sort Macropus is the most notable and incorporates a few famous animal groups, each with its own arrangement of qualities.

Red Kangaroo (Macropus rufus):

The Red Kangaroo is the biggest marsupial and the most famous agent of its loved ones. Known for its unmistakable ruddy earthy colored fur, the Red Kangaroo is a strong container, utilizing its solid rear appendages and long tail to proficiently cover huge distances.

Guys are for the most part bigger than females, and their vigorous form mirrors their predominance in friendly progressive systems.

The Red Kangaroo fundamentally possesses bone-dry and semi-dry areas, where its variations to the cruel environment incorporate proficient water preservation and thermoregulation. These kangaroos structure free gatherings known as crowds, and their ways of behaving, including boxing and pounding, assume key parts in correspondence and social elements.

Eastern Dim Kangaroo (Macropus giganteus):

The Eastern Dim Kangaroo is one more unmistakable individual from the Macropus variety. Dissimilar to the Red Kangaroo, it is portrayed by its delicate dim fur and an unmistakable lighter variety on its underparts. This species is tracked down in various natural surroundings, including meadows, forests, and waterfront regions.

Eastern Dim Kangaroos frequently display a more organized social association, shaping more modest, affectionate nuclear families. Strength among guys is laid out through size and age, and correspondence inside the gathering includes a scope of actual signals. These kangaroos are dominatingly herbivorous, benefiting from various grasses and bushes.

Western Dim Kangaroo (Macropus fuliginosus):

The Western Dim Kangaroo is a direct relation of the Eastern Dim, and the two species share numerous qualities. Notwithstanding, the Western Dim is for the most part more modest and has a hazier coat, going from brown to practically dark. This species occupies many conditions, including beach front regions, forests, and shrublands.

Like different kangaroos, Western Dark Kangaroos are herbivorous, depending on a tight eating routine of grasses and vegetation. Their social construction is portrayed by free gatherings, and guys might take part in fights to lay out strength. The capacity to cover enormous distances and adjust to various environments adds to the outcome of this species.

Antilopine Kangaroo (Macropus antilopinus):

The Antilopine Kangaroo is tracked down in northern Australia, occupying fields, savannas, and open forests. This species is perceived for its unmistakable appearance, with a light-hued face and body, diverging from hazier appendages. Antilopine Kangaroos are adjusted to a heat and humidity, where they face various difficulties contrasted with their bone-dry dwelling partners.

Socially, Antilopine Kangaroos structure bigger gatherings, and their progressive construction is kept up with through different ways of behaving, including vocalizations and actual showcases.

Their eating routine comprises predominantly of grasses, and their capacity to flourish in assorted environments features the versatility of kangaroo species.

Red-necked Wallaby (Macropus rufogriseus):

While in fact a wallaby, the Red-necked Wallaby is firmly connected with kangaroos and is worth focusing on for its remarkable qualities. Found in different living spaces, including timberlands and fields, this species is perceived by its rosy earthy colored fur and particular white cheek stripe.

Red-necked Wallabies are for the most part lone, and their conceptive methodology includes pocket based advancement like kangaroos. These wallabies are herbivores, benefiting from grasses, spices, and bushes. Their more modest size and nimbleness permit them to explore thick vegetation in forested conditions.

Rock Wallabies: Petrogale Class

Aside from the Macropus class, the family Macropodidae incorporates different genera, each with its own arrangement of kangaroo species. The family Petrogale involves rock wallabies, adjusted to rough and rough landscapes.

Yellow-footed Rock Wallaby (Petrogale xanthopus):

The Yellow-footed Rock Wallaby is a striking animal types known for its particular yellow feet and tail. Possessing rough outcrops and slopes, this wallaby is all around adjusted to exploring testing territory. The tinge of its fur gives viable disguise against the rough scenery.

These stone wallabies are herbivores, consuming different plants adjusted to dry conditions. Preservation endeavors are pivotal for the endurance of the Yellow-footed Rock Wallaby, as they face dangers from territory misfortune and presented hunters.

Unified Rock Wallaby (Petrogale assimilis):

The Unified Stone Wallaby, tracked down in northern Australia, is portrayed by its dull earthy colored fur and unmistakable facial markings. This species possesses a scope of conditions, including rough regions, woods, and beach front locales. Unified Rock Wallabies are adjusted to both earthly and arboreal movement.

Socially, these wallabies structure little gatherings, and their collaborations include a mix of vocalizations and actual showcases. Protection worries for the Partnered Rock Wallaby incorporate natural surroundings discontinuity and predation by presented species.

Rabbit Wallabies: Onychogalea Class

The Onychogalea variety incorporates bunny wallabies, more modest family members of kangaroos with particular qualities.

Harnessed Nailtail Wallaby (Onychogalea fraenata):

The Harnessed Nailtail Wallaby is a basically jeopardized animal types known for the unmistakable white "harness" checking on its shoulders. Generally boundless, natural surroundings misfortune and predation prompted an emotional decrease in their populace. Protection endeavors are progressing to guarantee the endurance of this one of a kind wallaby.

Harnessed Nailtail Wallabies are herbivores, benefiting from various grasses and vegetation. Endeavors to safeguard and reestablish their regular territory are basic for their recuperation.

Northern Nailtail Wallaby (Onychogalea unguifera):

The Northern Nailtail Wallaby is one more individual from the Onychogalea sort, tracked down in northern Australia. Its name is gotten from the particular prod like development on its tail. This wallaby is adjusted to different territories, including prairies and open forests.

Preservation status differs among populaces, and endeavors to screen and safeguard their environments are significant. Like other rabbit wallabies, the Northern Nailtail Wallaby faces dangers from environment misfortune and presented hunters.

1.2Unique physical and behavioral characteristics

Kangaroos, the notable marsupials of Australia, are prestigious for their particular physical and social qualities that have advanced more than huge number of years. From their strong rear appendages and jumping headway to their perplexing social designs and conceptive techniques, kangaroos stand apart as momentous instances of variation to their current circumstance. This investigation will dive into the one of a kind physical and social qualities that characterize these intriguing animals.

Exceptional Actual Qualities

1. **Jumping Headway:**
   The most famous component of kangaroos is without a doubt their unmistakable jumping motion. This particular type of development, known as saltatory movement, is extraordinary among warm blooded creatures and is a main trait of kangaroos. The strong rear appendages, adjusted for bouncing, empower them to cover tremendous distances with unbelievable speed and productivity.
   The construction of the rear appendages is a wonder of biomechanical plan. The lengthened and solid rear legs, joined with an abbreviated forelimb structure, are upgraded for drive during bouncing. The rear feet are huge and outfitted areas of strength for with, ligaments that behave like springs, putting away and delivering energy with each bounce. This energy-productive method of movement is critical for kangaroos, permitting them to cross different scenes, from open fields to thick vegetation.
2. **Tail as an Adjusting Organ:**
   One more novel actual quality of kangaroos is their long and strong tail, which serves numerous capabilities. While not prehensile, the tail assumes a pivotal part in keeping up with balance during quick jumps and sharp turns. Kangaroos utilize their tails as an offset, permitting them to explore fluctuated territories with nimbleness.
   The tail is likewise utilized as a strong propulsive organ during sluggish developments, like preparing or brushing. Its solid muscular structure offers extra help when kangaroos are fixed or participated in exercises that don't include jumping. This double usefulness features the flexibility of the kangaroo's life structures to various parts of its life.
3. **Sexual Dimorphism:**
   Kangaroos show sexual dimorphism, with guys for the most part being bigger and more strong than females. This size contrast is especially observable in bigger species like the Red Kangaroo, where grown-up guys can remain north of six feet tall and weigh in excess of 200 pounds, while females are for the most

part more modest and lighter.
This dimorphism isn't simply connected with size yet additionally reaches out to other actual elements. Guys frequently have more vigorous and strong forms, with bigger forelimbs that are utilized in ritualized battle during the reproducing season. The distinctions in size and morphology mirror the jobs every orientation plays in the social elements and conceptive procedures of kangaroo populaces.

4. **Dental Variations:**

Kangaroos are herbivores, and their dental variations mirror their particular eating routine, which comprises basically of extreme grasses and vegetation. The dental recipe of kangaroos incorporates sharp incisors at the front for editing grass, trailed by huge molars at the back for crushing and handling sinewy plant material.

The steady wear and regrowth of teeth all through their lives are critical for kangaroos to keep up with productive rumination. This dental variation permits them to separate greatest nourishment from the difficult plant material they consume, adding to their prosperity as herbivores in different biological systems.

Remarkable Conduct Qualities

1. **Marsupial Proliferation:**
   One of the most unmistakable conduct qualities of kangaroos is their method of proliferation as marsupials. Not at all like placental well evolved creatures, kangaroos bring forth moderately lacking live youthful, known as joeys. The female kangaroo has two uteri and the capacity to defer the improvement of the undeveloped organism until ecological circumstances are positive.
   After a short incubation period, the little, bare joey is conceived and naturally creeps into the mother's pocket, where it connects itself to a nipple for sustenance. The pocket gives a solid climate to the joey's initial turn of events, shielding it from outer dangers. The course of pocket based improvement is an unmistakable element of marsupials, permitting kangaroos to adjust to the unusual and frequently unforgiving states of their Australian territories.
2. **Social Construction and Ordered progression:**
   Kangaroos display different social designs and ordered progressions, shifting among species and impacted by variables like environment and food accessibility. The Red Kangaroo, for instance, frequently shapes free gatherings known as crowds. These hordes can comprise of the two guys and females, with a predominant male declaring command over assets and rearing open doors.
   The social construction of kangaroo populaces is frequently various leveled, with predominant people appreciating special admittance to mates and assets. Predominance is commonly settled through ritualized battle, where guys utilize their forelimbs to box with rivals, displaying their solidarity and assurance.

Conversely, species like the Eastern Dark Kangaroo structure more modest, all the more closely knit nuclear families. These gatherings are driven by a predominant male, and social elements inside the nuclear family are more mind boggling. The predominant male safeguards the gathering against outer dangers as well as decides admittance to assets and rearing honors.

3. **Specialized Techniques:**

   Correspondence among kangaroos is prevalently non-vocal, depending on a scope of actual signs and ways of behaving. Kangaroos use these specialized strategies to pass on data about their goals, societal position, and likely dangers. Some normal specialized techniques include:

   Body Stance: Kangaroos use body stance to pass on a scope of messages. Prevailing people might embrace more upstanding stances to declare their predominance, while agreeable people might squat or incline forward.

   Tail Position: The place of the tail is an essential specialized device. A raised tail might show hostility or energy, while a brought down tail can mean accommodation or a quiet state.

   Looks: Kangaroos utilize looks to convey feelings and aims. Extending the eyes, causing a stir, or uncovering the teeth can impart different messages.

   Pounding: Pounding the ground with the rear feet is a typical type of correspondence among kangaroos. This conduct effectively makes others aware of likely dangers and is in many cases utilized in light of seen peril.

   Boxing: Boxing is an actual type of correspondence, principally saw among male kangaroos during the reproducing season. It includes utilizing the forelimbs to strike an adversary, laying out predominance and settling clashes over mates and assets.

4. **Nighttime and Crepuscular Way of behaving:**

   Most kangaroo species are crepuscular, meaning they are generally dynamic during first light and nightfall. This social variation permits kangaroos to stay away from the intensity of the day, which is especially worthwhile in the parched and semi-dry conditions they frequently possess. During the day, kangaroos might rest in concealed regions, preserving energy for their more dynamic periods.

   The crepuscular way of behaving of kangaroos is likewise vital regarding hunter evasion. By being dynamic during low-light circumstances, kangaroos diminish their weakness to predation, as a considerable lot of their normal hunters are more dynamic during sunlight hours.

5. **Pounding and Boxing:**

   Two eminent social characteristics of kangaroos are pounding and boxing. Pounding, as referenced prior, includes musically hitting the ground with the rear feet. Kangaroos utilize this way of behaving to speak with different individuals from their gathering, making them aware of possible risks or flagging their presence.

Boxing, then again, is an actual presentation basically seen among male kangaroos during the rearing season. It is a ritualized type of battle where guys stand on their rear legs and utilize their forelimbs to strike rivals. Boxing is a method for laying out predominance inside the gathering and seeking admittance to mates.

6. **Specialty Variations:**

Different kangaroo species have adjusted to many specialties inside the Australian scene. For example, rock wallabies, like the Dark flanked Rock Wallaby, are capable climbers, using serious areas of strength for them appendages and tails to explore rough territory. Their specialty variation permits them to get to food and safe house in regions where different kangaroos could battle.

Essentially, species like the Antilopine Kangaroo have adjusted to tropical fields and savannas, displaying transient conduct looking for ideal taking care of grounds. These specialty transformations feature the flexibility of kangaroos in colonizing different living spaces and taking advantage of an assortment of food assets.

1.3Adaptations to the Australian environment

Australia, with its immense and different scenes, presents a novel arrangement of difficulties and valuable open doors for its local widely varied vegetation. North of millions of years, the life forms occupying this landmass have gone through wonderful variations to make due and flourish in a scope of biological systems, from parched deserts to lavish rainforests. This investigation digs into the phenomenal variations of Australian verdure to the unmistakable natural states of the landmass.

Vegetation Variations:

****1. Dry season Obstruction and Water Stockpiling:**

The bone-dry and semi-dry districts of Australia present critical provokes for vegetation because of restricted water accessibility. Numerous Australian plants have created shrewd transformations to adapt to delayed times of dry spell. One outstanding transformation is the improvement of profound roots that tap into underground water sources. Acacia species, generally known as wattles, represent this transformation, for certain species having roots that broaden a few meters underneath the surface.

Furthermore, deliciousness is a system utilized by certain plants to store water in their tissues. The notable delicious plant, the "Sturt's Desert Pea" (Swainsona formosa), found in bone-dry districts, stores water in its meaty stems and leaves. This permits the plant to get through expanded times of water shortage.

****2. Fire Transformations:**

Fires are a characteristic and intermittent element of the Australian scene. Numerous local plants have advanced variations to flourish in fire-inclined conditions. A few Australian plants, similar to the eucalypts, have grown thick, heat proof bark that safeguards them during fierce blazes. Furthermore, a few animal types display a peculiarity known as "fire-invigorated germination," where the intensity of a fire sets off the arrival of seeds from woody cases, advancing new development in the post-fire climate.

The "Banksia" class gives one more illustration of fire-adjusted greenery. Banksias have woody cones that delivery seeds because of fire, guaranteeing that the cutting edge can secure itself in the supplement rich debris beds left after a fire. This variation is critical for the recovery of plant networks in fire-inclined biological systems.

3. Xerophyte Variations:

Xerophytes are plants that have advanced to get by in dry or bone-dry circumstances. The Australian scene, especially in the inside, highlights broad parched districts where xerophytic transformations are predominant. One such variation is the presence of little, limited leaves or needle-like foliage, which limits water misfortune through happening.

The "Spinifex" grasses, prevailing in bone-dry districts, embody xerophyte variations with their extreme, wiry leaves that lessen water misfortune. Also, a few Australian plants, similar to the "Mulga" (Acacia aneura), have changed leaves known as phyllodes, which lessen the surface region presented to the sun, in this way preserving water.

4. Myrmecochory:

Myrmecochory, or seed dispersal by subterranean insects, is a captivating transformation found in a few Australian plants. Certain plant species produce seeds with a greasy member called an elaiosome, which draws in subterranean insects. The insects convey the seeds back to their homes, consume the elaiosome, and dispose of the seed in supplement rich insect hills. This cycle benefits both the plant and the insects, as the seeds get a supplement help in an ideal climate.

The "Acacia" family frequently displays myrmecochory, depending on insects for seed dispersal. This mutualistic connection among plants and insects upgrades the possibilities of fruitful germination and foundation, particularly in supplement unfortunate soils.

Fauna Variations:

1. Marsupial Generation:

One of the most particular highlights of Australian fauna is the commonness of marsupials, a gathering of vertebrates portrayed by bringing forth generally lacking live youthful, which then, at that point, keep on creating in a pocket. This extraordinary conceptive technique is a variation to the erratic ecological states of Australia.

Kangaroos, wallabies, and koalas are famous instances of marsupials in Australia. The pocket gives a protected and safeguarded climate for the lacking youthful, guaranteeing their endurance in a difficult and frequently asset restricted scene. This variation permits marsupials to be more adaptable in their conceptive timing, as they can defer the improvement of undeveloped organisms until conditions are great.

2. Nighttime and Crepuscular Way of behaving:

Numerous Australian creatures have adjusted their way of behaving to adapt to the outrageous temperatures of the day, particularly in bone-dry areas. Nighttime and crepuscular movement designs are normal among various species, permitting them to keep away from the intensity and decrease water misfortune through evaporative cooling.

The "Bilby" (Macrotis lagotis), a little, nighttime marsupial, is an illustration of an Australian creature with a basically nighttime way of life. By searching during the cooler evenings, the Bilby saves energy and water, improving its possibilities of endurance in the bone-dry conditions it possesses.

3. Tunneling and Aestivation:

Tunneling conduct is far reaching among Australian fauna, filling different needs like safe house from hunters, security from outrageous temperatures, and admittance to underground food sources. The "Bilby" is known for its proficient tunneling skills, making mind boggling frameworks of tunnels in the desert sand.

As well as tunneling, a few Australian creatures have adjusted to endure broadened times of outrageous intensity through aestivation. Aestivation includes a condition of torpidity or latency during hot and dry periods, assisting creatures with rationing energy and water. The "Fat-followed Dunnart" (Sminthopsis crassicaudata), a little marsupial, is known to enter a condition of aestivation during hot and dry circumstances.

4. Cover and Mimicry:

Cover and mimicry are normal variations in Australian fauna, permitting creatures to mix into their environmental factors or impersonate different species for different purposes, including staying away from hunters or ambushing prey.

The "Leaf-followed Gecko" (Saltuarius spp.) is an illustration of an Australian reptile that utilizes disguise. With its leveled body and leaf-like appearance, it can actually mix into the encompassing vegetation, making it less apparent to hunters and prey the same.

Mimicry is exhibited by the "Macleay's Swallowtail Butterfly" (Graphium macleayanus), which impersonates the presence of the harmful "Normal Crow Butterfly" (Euploea center). This type of mimicry, known as Batesian mimicry, gives security to the non-poisonous species by tricking likely hunters into thinking it is unsafe.

5. Movement and Nomadism:

A few Australian bird animal types have developed transient or traveling ways of behaving to adapt to changing ecological circumstances, especially because of occasional varieties in asset accessibility.

The "Rainbow Honey bee eater" (Merops ornatus) is a transient bird that traversed Australia, following the sprouting of blossoming eucalypts and the ensuing expansion in bug prey. Roaming conduct is seen in the "Budgerigar" (Melopsittacus undulatus), which can cover tremendous distances looking for food and water, showing versatility in light of flighty asset circulation.

Chapter 2

History of Kangaroos in Captivity

The enamoring charm of kangaroos, those notable marsupials local to Australia, has provoked people to bring them into bondage for quite a long time. From early European travelers exhibiting these one of a kind animals as extraordinary interests to the cutting edge time of natural life preservation and instructive drives, the historical backdrop of kangaroos in imprisonment is rich and various. This extensive investigation will disentangle the complex excursion of kangaroos in imprisonment, diving into the evolving discernments, inspirations, and works on encompassing the keeping of these amazing animals.

Early Experiences and Interests

1. **European Investigation and the Principal Kangaroos in Bondage**
   The historical backdrop of kangaroos in imprisonment can be followed back to the late eighteenth century when European pilgrims and naturalists originally set foot on the shores of Australia. The experience with the extraordinary fauna of the landmass, including kangaroos, ignited monstrous interest and interest among the pioneers.In 1770, during Chief James Cook's campaign, a kangaroo was shot and welcomed on board the HMS Try. This obvious one of the earliest examples of a kangaroo being taken into imprisonment, but momentarily. The interest encompassing these extraordinary animals developed, and examples were gathered for logical review and public presentation.
2. **Early Zoos and Zoological displays**

The foundation of zoological nurseries and zoos in the nineteenth century assumed an essential part throughout the entire existence of kangaroos in imprisonment. These establishments, frequently connected with imperial courts or wealthy people, intended to exhibit the variety of the world's fauna. Kangaroos, with their extraordinary appearance and bouncing motion, became star attractions in these early zoos.

One striking model is the London Zoo, which presented its most memorable kangaroo display in the mid 1800s. The kangaroos were shown close by other fascinating creatures, making an environment of miracle and shock. In any case, the circumstances in these early zoos were frequently distant from ideal, lacking appropriate comprehension of the dietary and natural necessities of kangaroos.

From Abuse to Protection

1. **Kangaroos in Bazaars and Diversion**
   In the late nineteenth and mid twentieth hundreds of years, the charm of kangaroos reached out to media outlets. Kangaroos were highlighted in carnivals, voyaging zoological displays, and event congregations all over the planet. These scenes benefited from the oddity and display of kangaroos, frequently exposing them to unnatural and distressing circumstances.
   The double-dealing of kangaroos for diversion purposes raised moral worries, and public opinion started to move towards a more preservation situated approach. As mindfulness developed in regards to the effect of bondage on creature government assistance, endeavors were started to work on the everyday environments of kangaroos in imprisonment.
2. **The Advancement of Zoos and Untamed life Preservation**

The mid-twentieth century saw a change in outlook in the way of thinking of zoos, progressing from simple grandstands of extraordinary creatures to organizations devoted to preservation, training, and examination. This change significantly affected the historical backdrop of kangaroos in imprisonment.

Zoos started to focus on the prosperity and preservation of species, including kangaroos. Walled in areas were intended to impersonate normal territories, and endeavors were made to lay out rearing projects to guarantee the endurance of jeopardized species. Kangaroos became diplomats for their wild partners, bringing issues to light about the dangers they looked in the wild, like territory misfortune and hunting.

The Difficulties of Hostage Reproducing

1. **Regenerative Difficulties**
   Hostage rearing projects for kangaroos confronted interesting difficulties because of the marsupial conceptive framework. Dissimilar to placental warm blooded creatures, kangaroos bring forth moderately lacking youthful, which then keep on creating in the pocket. Recreating the complicated conceptive cycles of kangaroos in imprisonment ended up being a mind boggling task.
   Rearing outcome in bondage was blocked by elements like pressure, fake conditions, and the shortfall of normal signals for propagation. Zoos and natural life foundations needed to concentrate on understanding the regenerative science of kangaroos to upgrade reproducing achievement.

2. **Spearheading Endeavors in Hostage Reproducing**

Regardless of the difficulties, a few zoos and untamed life associations spearheaded fruitful hostage rearing projects for kangaroos. The Taronga Zoo in Sydney, Australia, assumed a significant part in propelling procedures for reproducing kangaroos in bondage. Their endeavors incorporated the advancement of specific nooks, dietary plans, and rearing conventions.

These drives contributed not exclusively to the protection of kangaroo species yet additionally to the logical comprehension of marsupial propagation. The information acquired from hostage reproducing programs became important for analysts and protectionists pursuing the safeguarding of kangaroo populaces in nature.

Instructive Drives and Preservation Effort

1. **Kangaroos as Instructive Representatives**
   As zoos embraced a more preservation centered mission, kangaroos arose as strong instructive ministers. They became focal figures in instructive projects pointed toward bringing issues to light about Australian natural life, biological systems, and protection challenges. Zoos started to focus on interpretive showcases and intuitive encounters to draw in guests in finding out about kangaroos and their part in the climate.
   Instructive effort reached out past zoos, with untamed life safe-havens and nature saves integrating kangaroo displays into their projects. Intuitive experiences, directed visits, and instructive materials became fundamental apparatuses for conveying the meaning of kangaroo protection.
2. **Preservation Difficulties and Drives**

The preservation challenges looked by kangaroos in the wild, for example, natural surroundings misfortune, environmental change, and human-untamed life struggle, provoked expanded joint effort between hostage offices, protection associations, and government organizations. Hostage populaces filled in as repositories for hereditary variety, giving a likely source to renewed introduction projects and species recuperation endeavors.

A few kangaroo animal varieties, including the Red Kangaroo and the Western Dark Kangaroo, became central focuses for preservation drives. Preservation reproducing programs meant to keep up with hereditarily different populaces, guaranteeing the drawn out practicality of these species. Also, research led on hostage kangaroos contributed significant bits of knowledge into their way of behaving, nature, and well-being, helping preservation systems in nature.

Challenges and Moral Contemplations

1. **Moral Worries in Bondage**
   The historical backdrop of kangaroos in bondage isn't without its moral discussions. The constrainment of wild creatures, including kangaroos, raises moral worries connected with creature government assistance, social limitations, and the mental prosperity of people. The discussion encompassing the morals of keeping kangaroos in imprisonment keeps on developing, provoking a reconsideration of norms and practices inside the hostage climate.
   Endeavors have been made to address moral contemplations by carrying out enhancement programs, giving open and naturalistic fenced in areas, and accentuating the significance of species-explicit consideration. The moral talk encompassing hostage kangaroos highlights the continuous obligation to offset protection goals with the government assistance of individual creatures.
2. **Legitimate Securities and Guidelines**

The moral contemplations related with keeping kangaroos in bondage have prompted the foundation of lawful assurances and guidelines. Numerous nations, including Australia, have executed severe rules for the consideration and the board of hostage kangaroos. These guidelines incorporate walled in area principles, veterinary consideration, rearing practices, and protection objectives.

The lawful system encompassing hostage kangaroos looks to find some kind of harmony between protection, schooling, and creature government assistance. Specialists routinely screen offices to guarantee consistence with guidelines, and resistance might bring about punishments or the denial of licenses.

Future Bearings: Protection, Exploration, and Government assistance

1. **Incorporating Preservation and Exploration**
   The eventual fate of kangaroos in imprisonment is naturally connected to the continuous endeavors to monitor and research these famous marsupials. Propels in innovation, for example, hereditary sequencing and regenerative advances, hold guarantee for improving hostage rearing projects. Cooperative exploration projects between zoos, colleges, and preservation associations mean to address basic holes in how we might interpret kangaroo science, conduct, and wellbeing.
   Hostage populaces keep on assuming an essential part in supporting protection drives, especially for species confronting dangers in nature. Renewed introduction programs, living space rebuilding endeavors, and local area commitment projects are necessary parts of all encompassing preservation systems that influence the qualities of both hostage and wild populaces.
2. **Progresses in Creature Government assistance**

The development of creature government assistance principles stays a critical concentration for the eventual fate of kangaroos in bondage. Progressing examination into the conduct environment of kangaroos illuminates the plan regarding fenced in areas that take care of their regular ways of behaving, social designs, and exercise needs. Enhancement programs, dietary headways, and veterinary consideration conventions add to the general prosperity of kangaroos in bondage.

Public mindfulness and schooling efforts likewise assume a vital part in molding the eventual fate of kangaroos in bondage. Drawing in the general population in the protection story of kangaroos cultivates a feeling of obligation and energizes support for drives that focus on the government assistance and preservation of these notorious marsupials.

2.1Early attempts at kangaroo captivity

The early endeavors at kangaroo imprisonment are woven into the texture of Australia's pilgrim history, set apart by the experiences of European travelers with the landmass' exceptional and different natural life. From the underlying surprise and interest ignited by these impossible to miss marsupials to the difficulties of moving and keeping up with them in bondage, this account investigates the early sections of bringing kangaroos into human consideration.

Native Points of view

Prior to diving into the European investigation period, recognizing the long-standing association among kangaroos and Australia's Native peoples is fundamental. Native people group had existed together with kangaroos for millennia, incorporating them into their social practices, stories, and reasonable hunting rehearses. Kangaroos were a wellspring of food as well as held otherworldly importance, and their collaborations with Native people group molded an agreeable relationship with the land.

European Investigation and Interest

1. **Early Experiences by European Adventurers**
   The late eighteenth century saw the appearance of European travelers on the shores of Australia, and the curiosity of the local fauna, including kangaroos, enthralled the minds of those early swashbucklers. The primary recorded experience with kangaroos by Europeans happened during Chief James Cook's undertaking in 1770.
   Cook's team noticed and associated with kangaroos, archiving their perceptions in diaries and representations. Be that as it may, at this stage, bringing kangaroos into bondage for logical review or public show had not yet flourished. The essential focal point of these early pilgrims was on reporting and grasping the recently discovered biodiversity of Australia.
2. **Catching and Showing Kangaroos in Europe**

As information about Australia's one of a kind fauna spread across Europe, a developing interest with kangaroos arose. Examples, including kangaroos, were gathered during resulting endeavors and shipped back to Europe. These examples were in many cases protected through taxidermy or as skeletal presentations, filling in as interests in normal history assortments and confidential cupboards of interests.

In the late eighteenth and mid nineteenth hundreds of years, live kangaroos started to show up in confidential zoological gardens and displays all through Europe. Rich people, gatherers, and exhibitors considered the kangaroo's uniqueness to be an expected draw for public crowds anxious to observe colorful animals from far off lands.

Difficulties of Early Kangaroo Imprisonment

1. **Transportation Difficulties**

   The transportation of live kangaroos from Australia to Europe presented huge difficulties during the early endeavors at kangaroo bondage. The long ocean journeys exposed the creatures to upsetting circumstances, restricted space, and insufficient arrangements. Kangaroos, adjusted to the immense and open scenes of Australia, battled in the bound spaces of boats.

   The physiological and conduct variations of kangaroos, including their special regenerative framework and jumping velocity, were not completely perceived at that point. Subsequently, endeavors to move kangaroos frequently brought about high death rates, making it a considerable test to lay out reasonable hostage populaces in Europe.
2. **Dietary and Natural Difficulties**

Understanding the dietary and natural necessities of kangaroos in imprisonment ended up being another obstacle. The eating routine of kangaroos in the wild comprises principally of grasses and vegetation, and reproducing these dietary prerequisites in bondage was testing. Lacking sustenance and unsatisfactory conditions prompted medical problems and diminished the life span of kangaroos in imprisonment during this period.

The absence of information about kangaroo conduct and environment inclinations further confounded their consideration. Early efforts to give fitting fenced in areas frequently missed the mark, as the regular ways of behaving of kangaroos, for example, jumping and scrounging, were limited in bound spaces.

Striking Endeavors and Early Zoos

1. **The London Zoo and Early Kangaroo Shows**

   The foundation of zoological nurseries in the nineteenth century denoted a defining moment throughout the entire existence of kangaroo bondage.

   The London Zoo, established in 1828, was one of the spearheading organizations that planned to grandstand the world's fauna to general society. Kangaroos were among the colorful creatures highlighted in early shows at the London Zoo.

   The kangaroo shows at the London Zoo were met with extraordinary public interest, as guests wondered about the unconventional bouncing step and remarkable life systems of these Australian marsupials. The fame of kangaroo displays added to the developing pattern of exhibiting colorful untamed life in zoos as a type of public diversion and training.

2. **Public Discernment and Sentimentality**

The presence of kangaroos in zoos caught the public's creative mind and powered a feeling of miracle and interest. Kangaroos, alongside other colorful creatures, became images of investigation and the strange miracles of far off lands. The media of the time frequently sensationalized these showcases, adding to a romanticized and once in a while wrong depiction of kangaroos and Australia's untamed life.

The attention was principally on the oddity of these animals, and the difficulties they looked in bondage, both physiological and mental, were not completely perceived or tended to. Early kangaroo displays were described by an accentuation on scene as opposed to the government assistance and protection contemplations that would come to characterize current zoos.

Social Effect and the Advancement of Imprisonment

1. **Logical Interest and Grouping**
   Past the public interest, the bondage of kangaroos had logical ramifications. European naturalists and researchers saw the amazing chance to concentrate on these one of a kind marsupials very close, prompting headways in the comprehension of kangaroo science, life systems, and conduct. Kangaroos became subjects of logical examination, adding to the more extensive arrangement of marsupials and their position in the set of all animals.
2. **Advancement of Hostage Practices**

As logical comprehension progressed, the states of kangaroo imprisonment slowly gotten to the next level. Zoos and zoological displays started to put resources into additional reasonable nooks, endeavoring to reproduce regular territories that considered a more normal articulation of kangaroo ways of behaving. Further developed slims down, veterinary consideration, and cultivation rehearses meant to address the difficulties related with early kangaroo bondage.

The development of hostage rehearses mirrored a shift from simple show to a more protection situated approach. As zoos progressively perceived their part in natural life protection and schooling, endeavors were made to adjust kangaroo imprisonment to moral contemplations and preservation objectives.

2.2Evolution of kangaroo care in zoos and sanctuaries

The development of kangaroo care in zoos and safe-havens remains as a demonstration of the changing mentalities towards natural life the executives, protection, and the moral contemplations related with keeping these famous marsupials in bondage. From the beginning of interest driven shows to the current spotlight on protection, exploration, and training, the excursion follows a way set apart by logical headways, further developed cultivation rehearses, and an extending comprehension of kangaroo conduct and government assistance.

Early Practices and Difficulties

1. **Early Shows and Zoological gardens**
   The beginning of kangaroo imprisonment were portrayed by displays and zoological gardens that exhibited intriguing natural life for public amusement. Kangaroos, with their exceptional bouncing motion and charming life systems, became star attractions in zoological gardens like the London Zoo, where they were highlighted in the nineteenth 100 years.
   During this period, the essential spotlight was on the curiosity of these Australian marsupials, and the states of imprisonment were frequently less than ideal. Walled in areas missing the mark on comprehension of the normal ways of behaving and dietary requirements of kangaroos, bringing about difficulties connected with wellbeing and prosperity.
2. **Transportation and Natural Variables**

Shipping kangaroos from their local Australia to zoos in different areas of the planet presented critical difficulties. The long ocean journeys exposed the creatures to push, restricted space, and deficient arrangements. The physiological and conduct variations of kangaroos, including their exceptional regenerative framework and jumping movement, were not completely perceived, adding to high death rates during transportation.

Lacking information on kangaroo conduct and territory inclinations prompted the production of walled in areas that didn't address the creatures' regular requirements. Early endeavors at kangaroo care were hampered by an absence of comprehension of their dietary necessities, social designs, and the significance of recreating their regular habitat.

Moving Standards: The Preservation Period

1. **Zoos as Preservation Focuses**
   The last 50% of the twentieth century denoted a change in outlook in the way of thinking of zoos. Establishments progressed from being simple features of fascinating creatures to becoming habitats for preservation, training, and exploration. This change significantly affected kangaroo care, adjusting it to more extensive protection objectives.
   As zoos perceived their likely job in species preservation, kangaroos became central focuses for reproducing programs pointed toward keeping up with hereditary variety and adding to the endurance of imperiled species. Zoos progressively

teamed up with one another, natural life offices, and protection associations to foster techniques for both hostage and wild kangaroo populaces.

2. **Preservation Rearing Projects**

   Preservation rearing projects arose as a foundation of current kangaroo care. These projects planned to guarantee the drawn out reasonability of kangaroo populaces, especially for species confronting dangers in nature. The reproducing of kangaroos in bondage turned out to be more essential, zeroing in on keeping up with hereditary variety and tending to explicit protection needs.

   The Western Bog Turtle (Pseudemydura umbrina), for instance, turned into the focal point of an effective protection reproducing program. Also, the Western Dim Kangaroo (Macropus fuliginosus) and the Red Kangaroo (Macropus rufus) were remembered for oversaw rearing drives to help their protection status.

3. **Propels in Regenerative Science**

The difficulties related with kangaroo proliferation in imprisonment prompted huge progressions in conceptive science. As specialists acquired a more profound comprehension of kangaroo conceptive physiology, helped regenerative innovations (Workmanship) were acquainted with upgrade reproducing achievement. Strategies, for example, planned impregnation and in vitro preparation were investigated to defeat obstructions connected with pressure and sub-par rearing circumstances.

The Taronga Western Fields Zoo in Dubbo, Australia, turned into a spearheading organization in kangaroo regenerative examination. The zoo's fruitful reproducing programs, upheld by logical examination, have contributed essentially to the protection endeavors for a few kangaroo animal varieties.

Government assistance Contemplations and Moral Practices

1. **Grasping Kangaroo Conduct and Needs**

   The developing comprehension of kangaroo conduct and environment has driven changes in their consideration. Scientists and guardians have dug into the intricacies of kangaroo social designs, specialized strategies, and dietary inclinations.

   This information has been instrumental in planning fenced in areas that take care of the normal ways of behaving of kangaroos, permitting them to communicate their species-explicit characteristics.

   Perceptions of wild kangaroo conduct, like brushing, bouncing, and social connections, have educated the creation regarding enhanced conditions in imprisonment. Zoos and asylums endeavor to furnish roomy nooks with highlights that empower active work, mental feeling, and social associations among kangaroos.

2. **Walled in area Plan and Naturalistic Conditions**
   Current kangaroo walled in areas mean to reproduce the regular living spaces of these marsupials. Huge open spaces, fluctuated geography, and vegetation that imitates their local climate add to the prosperity of kangaroos in imprisonment. These naturalistic conditions address the actual requirements of kangaroos as well as give open doors to mental excitement and conduct advancement.
   Foundations, for example, the Australia Zoo have set benchmarks in fenced in area configuration, making vivid displays that focus on the government assistance and solace of kangaroos. The accentuation on naturalistic conditions lines up with a more extensive obligation to moral practices and the advancement of creature government assistance.
3. **Improvement Projects**

Perceiving the significance of mental feeling for kangaroos, advancement programs have become indispensable to their consideration. Improvement exercises are intended to empower regular ways of behaving, critical thinking, and actual activity. Normal enhancement strategies incorporate the presentation of novel items, food riddles, and potential open doors for social associations.

The Melbourne Zoo, for example, carries out an assortment of enhancement exercises for its kangaroo populace. These exercises not just add to the prosperity of the kangaroos yet additionally connect with guests in grasping the regular ways of behaving and transformations of these marsupials.

Schooling and Public Mindfulness

1. **Kangaroos as Diplomats for Protection**
   Zoos and safe-havens influence their job as instructive establishments to raise public mindfulness about kangaroos and their protection status. Kangaroos are frequently situated as representatives for their species in the wild, assisting guests with interfacing with the difficulties looked by these marsupials in their local environments.
   Instructive projects, directed visits, and intelligent displays give guests experiences into the normal history, conduct, and preservation needs of kangaroos.
   Establishments perceive the capability of these magnetic marsupials to motivate compassion, cultivating a feeling of obligation for the protection of their wild partners.
2. **Preservation Informing and Coordinated efforts**

Preservation informing has turned into a basic part of kangaroo care in zoos and asylums. Establishments team up with natural life offices, analysts, and ecological associations to convey the significance of territory safeguarding, dependable land the executives, and manageable practices for kangaroo protection.

The Taronga Preservation Society Australia, for instance, takes part in local area outreach programs, featuring the interconnectedness of human activities and the prosperity of kangaroos in nature. By underscoring preservation messages, establishments intend to activate public help for drives that add to the endurance of kangaroo species.

Legitimate Assurances and Guidelines

1. **Administrative Systems for Hostage Kangaroo Care**
   The rising accentuation on moral practices, creature government assistance, and preservation has prompted the foundation of lawful securities and guidelines overseeing kangaroo care in bondage. Numerous nations, including Australia, have explicit rules illustrating the norms for the administration, lodging, and care of kangaroos in zoos and safe-havens.
   Guidelines cover angles, for example, walled in area plan, veterinary consideration, rearing projects, and record-keeping. Specialists consistently examine offices with guarantee consistence, and organizations should comply to these rules to keep up with grants for keeping kangaroos in imprisonment.
2. **Backing for Moral Norms**

Creature government assistance associations and promoters assume a pivotal part in considering establishments responsible for the moral treatment of kangaroos in bondage. Progressing support endeavors center around bringing issues to light about the government assistance requirements of kangaroos and advancing moral norms in their consideration.

Public investigation, upheld by logical proof and moral contemplations, has impacted the advancement of kangaroo care rehearses. Backing drives have provoked foundations to reconsider and further develop their cultivation strategies, adding to a more caring and mindful way to deal with kangaroo the board.

Future Headings: Coordinating Science, Government assistance, and Preservation

1. **Progresses in Exploration and Innovation**
   The eventual fate of kangaroo care in zoos and asylums is molded by continuous exploration and mechanical headways. The reconciliation of hereditary examination, conceptive advances, and social investigations holds guarantee for improving the prosperity and preservation results for kangaroos.
   Logical forward leaps in helped regenerative advancements, like in vitro treatment and cryopreservation of hereditary material, may add to the hereditary variety and versatility of hostage kangaroo populaces. Propels in following innovation and remote observing will additionally refine how we might interpret kangaroo conduct in bondage, working with more custom-made care draws near.

2. **All encompassing Preservation Methodologies**
   The advancing job of zoos and asylums in kangaroo preservation stretches out past hostage reproducing to comprehensive protection techniques. Organizations progressively add to living space safeguarding, biological exploration, and local area commitment drives that help kangaroo populaces in nature.
   Cooperative endeavors between hostage offices, protection associations, and government organizations will be fundamental for executing compelling preservation techniques. The reconciliation of hostage and wild kangaroo populaces into more extensive protection plans will reinforce the flexibility of these famous marsupials confronting different dangers, including natural surroundings misfortune and environmental change.
3. **Proceeded with Instruction and Public Commitment**

Schooling and public commitment will stay fundamental parts of kangaroo care from now on. Establishments will keep on assuming an essential part in cultivating an association among guests and kangaroos, motivating a feeling of obligation for their preservation.

Imaginative instructive projects, intuitive shows, and computerized stages will be utilized to contact assorted crowds and pass on protection messages. By encouraging a more profound comprehension of kangaroo science, conduct, and preservation challenges, foundations plan to develop a worldwide local area focused on the security of these meaningful marsupials.

2.3Ethical considerations and controversies

The imprisonment of kangaroos has been a subject of continuous moral contemplations and contentions, mirroring an intricate interchange between preservation objectives, creature government assistance concerns, and public perspectives.

As the job of zoos and safe-havens in kangaroo care has advanced, so too have the moral norms and discussions encompassing the imprisonment of these notable marsupials. This investigation digs into the multi-layered moral scene, inspecting the key contemplations, contentions, and the continuous endeavors to figure out some kind of harmony between preservation objectives and the prosperity of individual kangaroos.

Preservation Goals and Moral Problems

1. **Preservation as a Main thrust**
   One of the essential inspirations driving the imprisonment of kangaroos is the preservation basic. Zoos and safe-havens assume a critical part in reproducing programs pointed toward saving jeopardized species and keeping up with hereditary variety. By taking part in hostage reproducing drives, establishments add to the protecting of kangaroo populaces, which face dangers, for example, living space misfortune, environmental change, and human-natural life struggle.

Protection centered hostage reproducing programs expect to make repositories of hereditary variety that can be used for likely renewed introduction into the wild or to reinforce existing populaces. This approach lines up with the more extensive preservation objectives of guaranteeing the endurance and versatility of kangaroo species in their local living spaces.

2. **Moral Quandaries in Preservation Rearing**

While the preservation advantages of hostage reproducing are clear, moral issues emerge simultaneously. The hostage rearing of kangaroos includes the administration of people's regenerative lives, frequently in conditions that vary altogether from their regular environments. The test lies in adjusting the likely advantages to the species with the moral contemplations connected with individual government assistance, independence, and regular ways of behaving.

Questions emerge in regards to the innate worth of people in bondage and the compromises between the aggregate preservation objectives and the prosperity of individual kangaroos. As establishments explore these moral difficulties, discusses arise about the suitability of hostage reproducing programs and the effect on the physical and mental wellbeing of kangaroos.

Creature Government assistance and Moral Contemplations

1. **Imprisonment Initiated Pressure and Conduct Limitations**

   One of the focal moral contemplations in kangaroo care rotates around the possible pressure and conduct limitations forced by imprisonment. Kangaroos are adjusted to the tremendous and open scenes of Australia, and the imprisonment of these exceptionally portable creatures in nooks can prompt pressure and disappointment.

   Social limitations, including restricted space for bouncing and rummaging, may think twice about regular ways of behaving fundamental for the physical and mental prosperity of kangaroos.

   The mental effect of imprisonment on kangaroos is an area of continuous exploration and moral investigation. Social stereotypies, like tedious developments or pacing, may appear in hostage kangaroos as a reaction to stretch or natural requirements. Moral worries emerge when bondage instigated pressure compromises the personal satisfaction for individual kangaroos.

2. **Nook Plan and Ecological Improvement**

Addressing moral worries connected with imprisonment includes an emphasis on nook plan and ecological enhancement. Establishments endeavor to establish conditions that copy the regular territories of kangaroos, giving adequate space, changed geography, and vegetation. These endeavors expect to improve the physical and mental prosperity of kangaroos by permitting them to communicate their regular ways of behaving, take part in friendly connections, and experience mental excitement.

Moral contemplations stretch out to the arrangement of ecological advancement, for example, food puzzles, novel items, and open doors for social commitment. Improvement programs are intended to moderate fatigue and empower species-explicit ways of behaving, adding to a more enhanced and satisfying life for kangaroos in imprisonment.

Moral Contentions: Public Insight and Basic entitlements

1. **Public Insight and Profound Association**
   Public discernment assumes a critical part in the moral debates encompassing kangaroo care in imprisonment. Zoos and safe-havens, as open confronting organizations, are dependent upon examination and differing suppositions in regards to the morals of keeping kangaroos. The close to home association that individuals structure with these notable marsupials frequently shapes their viewpoints on the propriety of imprisonment.
   While certain people view zoos and safe-havens as fundamental supporters of preservation and instruction, others express worries about the effect of bondage on the prosperity of kangaroos. Moral debates emerge when public opinion impacts approaches, the board rehearses, and the needs of establishments really focusing on kangaroos.
2. **Basic entitlements Promotion**

The moral discussions encompassing kangaroo care meet with more extensive conversations about basic entitlements and the moral treatment of creatures in imprisonment.

Basic entitlements advocates contend that the actual demonstration of keeping wild creatures for human perception and amusement is innately unscrupulous. They battle that the interests and independence of individual creatures ought to overshadow aggregate preservation objectives.

Issues like the right to opportunity of development, articulation of normal ways of behaving, and evasion of imprisonment prompted pressure are key to basic entitlements viewpoints. Advocates call for more prominent straightforwardness, examination, and adherence to moral principles in kangaroo care, encouraging foundations to focus on the government assistance of individual creatures inside the more extensive setting of preservation.

Legitimate Assurances and Administrative Systems

1. **Legitimate Assurances for Hostage Kangaroos**
   Perceiving the moral contemplations related with kangaroo care, numerous nations, including Australia, have executed legitimate assurances and administrative

structures overseeing the administration of kangaroos in imprisonment. These guidelines envelop a scope of viewpoints, including nook plan, veterinary consideration, reproducing practices, and protection objectives.
The legitimate structure intends to lay out norms for the moral treatment of kangaroos, giving rules to establishments to adhere to. Specialists screen and authorize these guidelines through normal examinations, license prerequisites, and punishments for resistance. Legitimate securities add to a degree of oversight that lines up with moral contemplations for hostage kangaroo care.

2. **Challenges in Authorization and Oversight**

Regardless of lawful assurances, challenges continue in the authorization and oversight of kangaroo care in imprisonment. Restricted assets, fluctuating understandings of guidelines, and contrasts in authorization needs across purviews add to irregularities in the use of legitimate norms. Furthermore, the unique idea of moral contemplations requires a nonstop assessment of existing guidelines to guarantee they line up with developing viewpoints on creature government assistance and preservation.

Endeavors to upgrade lawful assurances frequently include cooperation between government offices, creature government assistance associations, and the foundations liable for kangaroo care. Finding some kind of harmony between the requirement for oversight and the adaptability to adjust to new logical experiences and moral guidelines is a continuous test in the administrative scene.

Future Headings: Incorporating Morals and Protection

1. **Straightforwardness and Public Commitment**
   The fate of moral kangaroo care includes a guarantee to straightforwardness and public commitment. Establishments perceive the significance of giving data about their kangaroo care rehearses, protection drives, and moral contemplations to general society. Straightforward correspondence encourages trust and empowers informed public talk on the moral elements of kangaroo imprisonment.
   Public commitment methodologies, like instructive projects, directed visits, and intelligent displays, assume an essential part in conveying the intricacies of kangaroo care. By including general society in conversations about preservation, creature government assistance, and moral practices, establishments expect to develop a mutual perspective of the difficulties and obligations related with really focusing on kangaroos in imprisonment.
2. **Moral Exploration and Headways**
   The coordination of moral contemplations into research rehearses is a critical part representing things to come of kangaroo care. Logical headways, especially in the fields of creature conduct, veterinary science, and regenerative advances, hold the possibility to improve the prosperity of kangaroos in bondage. Moral

examination systems, including painless checking procedures and conduct perceptions, add to a more profound comprehension of kangaroo science and illuminate more humane consideration rehearses.

Developments in helped regenerative advancements, like managed impregnation and in vitro preparation, ought to be directed by moral rules that focus on the government assistance of individual creatures. Finding some kind of harmony between logical headways and moral contemplations is fundamental for guaranteeing that kangaroo care rehearses line up with advancing principles.

3. **Worldwide Coordinated effort for Moral Guidelines**

The moral contemplations encompassing kangaroo care stretch out past public boundaries, requiring worldwide joint effort to lay out and maintain moral guidelines. Global participation among zoological associations, protection bodies, and creature government assistance promoters can add to the improvement of rules that mirror an agreement on the moral treatment of kangaroos in bondage.

Shared accepted procedures, research discoveries, and moral structures can upgrade the aggregate comprehension of kangaroo care. Cooperative endeavors can likewise address difficulties related with the transportation of kangaroos across borders, guaranteeing that moral contemplations are focused on in worldwide preservation and hostage reproducing drives.

Chapter 3

Captive Kangaroo Environments

Hostage kangaroo conditions address a basic point of interaction between the special environmental variations of these notable marsupials and the obligations of human guardians. This far reaching investigation dives into the multi-layered parts of hostage kangaroo conditions, enveloping plan contemplations, government assistance rehearses, protection objectives, and the advancing methodologies that shape the prosperity of kangaroos in imprisonment.

Grasping the Special Requirements of Kangaroos

1. **Variations to the Australian Climate**

Prior to diving into the particulars of hostage conditions, understanding the regular transformations of kangaroos to the Australian environment is fundamental. Kangaroos are appropriate to the tremendous and various scenes of Australia, with transformations that incorporate strong rear appendages for jumping, a particular stomach related framework for handling stringy vegetation, and an extraordinary regenerative methodology, including the pocket and early stage diapause.

These transformations have developed north of millions of years because of the difficulties presented by Australia's dry and variable environment. Reproducing these variations in imprisonment requires cautious thought of the physical, physiological, and social necessities of kangaroos.

Standards of Hostage Climate Plan

1. **Space and Fenced in area Plan**
   One of the major standards in planning hostage kangaroo conditions is giving more than adequate space that permits to normal ways of behaving, including bouncing and scavenging. Kangaroos are profoundly portable creatures, and keeping them to little spaces can prompt pressure, social issues, and compromised actual wellbeing.

Nook configuration ought to consider the social design of kangaroos, offering regions for people to withdraw, rest, and participate in friendly communications. Changed geology, like slopes and shakes, can add to the psychological and actual feeling of kangaroos, emulating the variety of their regular environments.

2. **Vegetation and Diet**
   Repeating the kangaroos' regular eating regimen is urgent for their prosperity in imprisonment. Hostage conditions ought to highlight an assortment of vegetation, including grasses, bushes, and peruse, to give a different and healthfully adjusted diet. Vegetation fills in as a food source as well as a fundamental part of the kangaroos' normal way of behaving, permitting them to brush, peruse, and participate in scrounging exercises.
   Conditions ought to be intended to empower normal taking care of ways of behaving, like brushing and specific perusing. The accessibility of local plants that reflect those found in their local territories adds to the mental prosperity of kangaroos.
3. **Asylum and Solace**

Giving asylum is critical to kangaroos in hostage conditions, offering security from outrageous weather patterns, including sun, downpour, and temperature variances. Concealed regions, havens, and designs that copy normal highlights add to the prosperity of kangaroos, permitting them to look for asylum when required.

Happy with resting regions, like delicate substrates or bedding materials, ought to be integrated into the plan to help the normal ways of behaving of resting and resting. Regard for these subtleties upgrades the general solace and government assistance of kangaroos in bondage.

Conduct Advancement and Mental Excitement

1. **Significance of Conduct Advancement**
   Conduct improvement is a basic part of hostage kangaroo conditions, tending to the mental prosperity of people. Enhancement exercises plan to animate normal ways of behaving, empower critical thinking, and give mental commitment. Enhancement isn't just about actual activity yet in addition about encouraging mental feeling and forestalling fatigue.
   The consideration of articles like logs, rocks, and novel things in the climate can act as advancement apparatuses. Puzzle feeders, hanging peruse, and other intuitive gadgets urge kangaroos to participate in exercises that reflect their normal ways of behaving, advancing physical and emotional wellness.
2. **Social Connections**

Kangaroos are social creatures with complex social designs. Hostage conditions ought to be intended to work with positive social associations among kangaroos.

Gathering people in light of similarity, age, and social elements is fundamental for advancing a feeling of local area and diminishing pressure.

Noticing regular social progressive systems and taking into account social holding adds to the prosperity of kangaroos. Overseers screen overall vibes, interceding when important to guarantee the wellbeing and agreement of the gathering.

Wellbeing and Veterinary Contemplations

1. **Veterinary Consideration and Observing**
   Hostage kangaroo conditions require cautious veterinary consideration and checking to guarantee the wellbeing and prosperity of people. Standard wellbeing checks, preventive consideration, and brief mediation in the event of wounds or diseases are necessary to keeping up with the general strength of kangaroos.
   Veterinary experts team up with guardians to lay out wellbeing conventions, lead routine assessments, and execute preventive measures, for example, immunizations and parasite control. Observing the kangaroos' state of being, conduct, and conceptive wellbeing adds to early discovery and mediation.
2. **Stress Decrease Techniques**

Stress decrease is a vital part of overseeing hostage kangaroo conditions. Stress can emerge from different variables, remembering changes for the climate, human cooperations, or disturbances in friendly elements. Guardians carry out pressure decrease methodologies, like steady acquaintances with new conditions, limiting unsettling influences, and giving natural articles or associates, to moderate possible stressors.

Noticing kangaroo conduct is essential for distinguishing indications of stress, and overseers utilize their aptitude to change the executives rehearses in like manner. Stress decrease measures add to the general government assistance and flexibility of kangaroos in bondage.

Regenerative Administration

1. **Regenerative Physiology of Kangaroos**
   Understanding the regenerative physiology of kangaroos is fundamental for viable conceptive administration in imprisonment. Female kangaroos have a special regenerative framework, with the capacity to enter a condition of early stage diapause, permitting them to postpone the improvement of the undeveloped organism until ecological circumstances are positive for the endurance of the posterity.
   Hostage conditions ought to think about the conceptive necessities of kangaroos, giving fitting circumstances to reproducing, birthing, and pocket advancement. Cautious observing of conceptive cycles and ways of behaving helps

guardians in upgrading regenerative accomplishment while focusing on the government assistance of the two females and their posterity.

2. **Helped Conceptive Innovations**

Headways in helped regenerative advances (Workmanship) assume a part in dealing with the conceptive soundness of kangaroos in bondage. Strategies, for example, managed impregnation and in vitro treatment add to rearing projects by addressing difficulties connected with pressure, contradictory pairings, or conceptive issues.

Overseers work in a joint effort with regenerative experts to execute Craftsmanship methodologies, guaranteeing that these advancements line up with moral contemplations and focus on the government assistance of individual kangaroos. Effective regenerative administration upholds hereditary variety and adds to the preservation objectives of hostage populaces.

Protection and Training Objectives

1. **Job of Hostage Conditions in Protection**
   Hostage kangaroo conditions assume a pivotal part in supporting preservation endeavors for different species. By partaking in all around oversaw hostage reproducing programs, organizations add to the conservation of hereditary variety and the possible renewed introduction of people into nature.
   Preservation centered hostage conditions act as supplies for species confronting dangers, for example, living space misfortune, environmental change, and human-natural life struggle. The information acquired from concentrating on kangaroos in bondage adds to more extensive protection methodologies, including natural surroundings safeguarding, environmental examination, and local area commitment.
2. **Instructive Projects and Effort**

Hostage conditions act as instructive center points, offering potential open doors for the general population to find out about kangaroos, their biological systems, and protection challenges. Instructive projects, directed visits, and intuitive shows give guests experiences into the normal ways of behaving, variations, and protection needs of kangaroos.

Establishments influence their job as instructive stages to bring issues to light about the significance of natural life preservation, mindful land the executives, and manageable practices. Drawing in people in general in discussions about kangaroo care encourages a feeling of association and obligation regarding the prosperity of these notorious marsupials.

Challenges and Moral Contemplations

1. **Moral Contemplations in Hostage Conditions**
   The plan and the board of hostage kangaroo conditions raise moral contemplations that require cautious consideration. Offsetting the preservation basic with the government assistance of individual kangaroos is a perplexing test.
   Moral contemplations include issues like the effect of imprisonment on regular ways of behaving, stress decrease, social elements, and the fittingness of rearing projects.
   Straightforwardness in correspondence, adherence to moral norms, and progressing endeavors to address government assistance concerns are fundamental parts of dependable kangaroo care. Moral contemplations are vital to dynamic cycles, requiring a guarantee to ceaseless improvement in view of logical experiences and moral standards.
2. **Challenges in Hostage Climate The board**

Overseers face different difficulties in overseeing hostage kangaroo conditions. These difficulties incorporate the counteraction of stress-related ways of behaving, medical problems, conceptive administration intricacies, and adjusting to the assorted requirements of various kangaroo species.

Viable administration requires a mix of logical information, functional experience, and a profound comprehension of kangaroo conduct. The powerful idea of hostage conditions requires continuous exploration, preparing, and joint effort among overseers, veterinarians, and protection specialists.

Mechanical Developments in Hostage Care

1. **Mechanical Advances in Observing**
   Mechanical developments assume a part in improving the consideration of kangaroos in bondage. Observing gadgets, like cameras and sensors, furnish guardians with ongoing data about kangaroo conduct, movement levels, and natural circumstances. Remote observing innovations add to the early location of medical problems and backing pressure decrease procedures.
   Headways in GPS beacons permit guardians to accumulate information on kangaroo development designs, social communications, and territory use. This data adds to the refinement of fenced in area plans and the board works on, adjusting hostage conditions to the regular ways of behaving of kangaroos.
2. **Computerized Stages for Instruction**

Advanced stages offer open doors for expanding the effect of hostage conditions past actual limits. Virtual visits, instructive applications, and online assets empower foundations to contact a worldwide crowd, sharing data about kangaroo care, protection drives, and the significance of biodiversity.

Intelligent computerized stages draw in crowds in virtual encounters, encouraging an association with kangaroos and advancing preservation mindfulness. Foundations influence innovation to improve instructive effort and supporter for moral principles in kangaroo care.

Future Bearings in Hostage Kangaroo Conditions

1. **Joining of Exploration and Preservation**
   The eventual fate of hostage kangaroo conditions includes a more profound combination of examination and preservation objectives. Organizations will keep on adding to logical information through examinations on conduct, regenerative physiology, and environmental connections. Research discoveries will illuminate hostage the executives practices and improve protection techniques for kangaroos in nature.
   Cooperative endeavors between hostage offices, preservation associations, and exploration foundations will drive headways in understanding kangaroo science and the more extensive environments they occupy. The incorporation of hostage and wild populaces into all encompassing protection plans will reinforce worldwide drives for kangaroo preservation.
2. **Public Commitment and Backing**
   The job of public commitment and backing will fill in importance, molding the fate of hostage kangaroo care. Foundations will zero in on cultivating a feeling of obligation and compassion among guests, empowering support for preservation endeavors and moral norms in kangaroo care.
   Inventive instructive projects, intelligent shows, and advanced stages will be utilized to impart the significance of kangaroo preservation. Public commitment will reach out past actual visits to hostage conditions, making a worldwide local area focused on the prosperity and preservation of kangaroos.
3. **Manageable Practices and Moral Guidelines**

The fate of hostage kangaroo conditions includes a promise to manageable practices and moral principles. Foundations will keep on developing their methodologies, focusing on the government assistance of individual kangaroos while adding to more extensive protection objectives. Maintainable practices envelop mindful rearing, natural surroundings safeguarding, and moral contemplations in hostage the board.

Cooperative endeavors inside the zoo and protection local area will drive the foundation of moral principles for kangaroo care. Progressing discourse, research drives, and public mindfulness missions will add to the improvement of best practices that line up with advancing moral contemplations.

3.1 Zoo enclosures vs. wildlife sanctuaries

The discussion between zoo walled in areas and untamed life safe-havens has been at the front of conversations in the domain of natural life preservation, schooling, and creature government assistance. The two organizations assume fundamental parts in the conservation of biodiversity, however they vary essentially in their goals, approaches, and the encounters they proposition to the two creatures and guests.

This investigation dives into the unmistakable attributes of zoo fenced in areas and untamed life safe-havens, analyzing their motivations, the board rehearses, moral contemplations, and the developing jobs they play in the more extensive scene of natural life preservation.

Zoo Fenced in areas: Protection, Training, and Public Commitment

1. **Preservation Command**
   Zoos, as foundations, have advanced after some time from being principally amusement scenes to turning out to be strong promoters for untamed life protection. Numerous cutting edge zoos are effectively associated with reproducing programs, research drives, and cooperative endeavors pointed toward safeguarding jeopardized species and keeping up with hereditary variety. The preservation order of zoos is many times reflected in their support in Species Endurance Plans (SSPs) and global protection associations.
   Zoo nooks are painstakingly intended to feature different species, furnishing guests with the chance to notice and find out about creatures from around the world. The presence of these living envoys effectively brings issues to light about the dangers looked by natural life, including environment misfortune, environmental change, and poaching.
2. **Instructive Open doors**
   Instruction is a focal mainstay of zoos, and very much planned nooks act as unique study halls. Interpretive signage, directed visits, and intuitive shows offer guests bits of knowledge into the regular ways of behaving, transformations, and preservation needs of the creatures in plain view. Zoos plan to rouse a feeling of marvel, interest, and compassion, encouraging an association among guests and the untamed life they experience.
   The hostage setting considers controlled and organized instructive projects, including creature showings, talks by untamed life specialists, and vivid displays that reproduce normal environments. These encounters are intended to convey complex environmental ideas, advance preservation mindfulness, and persuade guests to make a move on the side of natural life protection.
3. **Challenges and Moral Contemplations**

While zoos assume an essential part in preservation and schooling, they likewise face moral difficulties and examination. Concerns spin around the moral treatment of creatures in bondage, the effect of imprisonment on regular ways of behaving, and the potential pressure experienced by creatures in zoo conditions. Pundits contend that the bondage of wild creatures out there in the open brings up moral issues about the government assistance of individual creatures and their right to opportunity.

Zoos have answered these worries by focusing on creature government assistance, executing advancement projects, and putting resources into exploration to more readily comprehend and address the requirements of hostage species. Moral contemplations are essential to current zoo the board, and numerous foundations are effectively participated in endeavors to further develop nook plan, veterinary consideration, and by and large guidelines of creature government assistance.

Natural life Safe-havens: Giving Asylum and Recovery

1. **Shelter for Safeguarded and Restored Creatures**
   Natural life safe-havens, rather than zoos, center around giving asylum and restoration to creatures that have been saved from unfavorable circumstances. Numerous safe-haven inhabitants are creatures that have been stranded, harmed, or dislodged because of elements, for example, natural surroundings annihilation, unlawful untamed life exchange, or experiences with human exercises. The essential objective of asylums is to offer a protected and naturalistic climate where creatures can carry on with out their lives in conditions that look like their wild territories.
   Natural life safe-havens frequently team up with protection associations, policing, and restoration focuses to save and really focus on creatures that can't be delivered once again into nature. Safe-havens act as long lasting homes for creatures that might have physical or conduct difficulties that keep them from enduring freely in their common habitats.
2. **Hands-Off Way to deal with Creature Communication**
   Natural life safe-havens by and large embrace a hands-off way to deal with creature cooperation, underscoring negligible human contact and intercession. The emphasis is on giving a peaceful climate where creatures can communicate normal ways of behaving, structure social securities, and, when pertinent, participate in recovery processes. Not at all like zoos, safe-havens focus on the prosperity of individual creatures over open amusement.
   The restricted human contact in safe-havens plans to diminish pressure and guarantee that creatures keep a degree of ferocity. This approach lines up with the way of thinking that creatures in safe-havens are not in plain view for human entertainment yet rather exist in a safeguarded space where their requirements are the essential thought.
3. **Challenges and Moral Contemplations**

Untamed life safe-havens face difficulties connected with subsidizing, assets, and the intricacies of really focusing on creatures with differed chronicles and needs. The moral contemplations for safe-havens rotate around guaranteeing the government assistance of individual creatures, giving fitting veterinary consideration, and tending to the drawn out monetary maintainability of the asylum.

The non-business nature of safe-havens, combined with their obligation to creature government assistance, recognizes them from conventional zoos. Moral contemplations incorporate keeping up with the physical and emotional wellness of asylum occupants, supporting normal ways of behaving, and establishing conditions that focus on the prosperity of creatures over open showcase.

Advancing Viewpoints: Crossover Models and Preservation Drives

1. **Mixture Models: Joining Protection and Government assistance**
   Perceiving the qualities and difficulties of both zoo nooks and natural life safe-havens, a few establishments have taken on crossover models that plan to find some kind of harmony between protection, training, and creature government assistance. These models look to give vivid encounters to guests while focusing on the prosperity and regular ways of behaving of creatures.
   Half and half offices frequently include extensive fenced in areas with naturalistic components, advancement programs, and instructive parts. The accentuation is on establishing conditions that take care of the necessities of the two creatures and guests, encouraging a more profound comprehension of natural life and advancing preservation mindfulness.
2. **Protection Drives: Overcoming any issues**

A shared conviction among zoos and safe-havens is their common obligation to natural life preservation. Numerous zoos effectively add to worldwide protection drives, supporting field projects, living space safeguarding, and research attempts. By overcoming any barrier among hostage and wild populaces, zoos add to the more extensive objective of shielding biodiversity and tending to the difficulties looked by species in their regular environments.

Joint efforts between zoos, asylums, and preservation associations are progressively normal, featuring the interconnectedness of endeavors to safeguard untamed life. These organizations perceive the reciprocal jobs that various foundations play in resolving the perplexing and complex issues influencing worldwide biodiversity.

3.2Design considerations for kangaroo habitats

Making territories for kangaroos in bondage requires cautious thought of their exceptional physiological, conduct, and natural necessities. Kangaroos, adjusted to the far reaching scenes of Australia, have unmistakable necessities that should be addressed in fenced in area plan to advance their prosperity. This investigation frames key plan contemplations for kangaroo territories, stressing the significance of reproducing regular ways of behaving, guaranteeing actual wellbeing, and supporting preservation and instructive objectives.

1. **Satisfactory Room and Geography**
   One of the principal contemplations in kangaroo natural surroundings configuration is giving more than adequate space and changed geography. Kangaroos are famous for their strong rear appendages and particular bouncing movement, which requires adequate room for exercise and normal development. Nooks ought to be sufficiently extensive to permit kangaroos to participate in bouncing, brushing, and social communications.
   Fluctuated geology, like slopes, shakes, and open spaces, mirrors the variety of Australia's scenes. The incorporation of these components improves the actual activity of kangaroos as well as gives open doors to mental feeling and investigation. All around planned environments consider the requirement for both open regions and segregated spaces where kangaroos can rest and withdraw.
2. **Vegetation and Scavenging Amazing open doors**
   Imitating the regular eating routine of kangaroos is vital for their wellbeing and prosperity. Kangaroos are herbivores that nibble on different grasses, bushes, and peruse. Thusly, kangaroo living spaces ought to highlight a different scope of vegetation, including local grasses and plants that reflect those tracked down in their common habitats.
   Planning environments with more than adequate scrounging open doors empowers normal taking care of ways of behaving and upholds the dietary requirements of kangaroos. Counting eatable vegetation and peruse permits kangaroos to communicate their instinctual ways of behaving of brushing and specific perusing, adding to their in general physical and emotional wellness.
3. **Cover and Ecological Enhancement**
   Giving safe house is fundamental in kangaroo natural surroundings to offer assurance from outrageous atmospheric conditions, including sun, downpour, and temperature vacillations. While kangaroos are adjusted to outside residing, havens ought to be decisively positioned to permit them to look for shelter when required.
   Ecological improvement is one more basic part of kangaroo living space plan. Enhancement exercises, for example, giving novel items, puzzle feeders, and designs that empower investigation, add to mental feeling and forestall weariness. Improvement programs mean to duplicate the difficulties of the wild, cultivating regular ways of behaving and upgrading the general prosperity of kangaroos in bondage.
4. **Social Elements and Gathering Systems**
   Kangaroos are social creatures with complex social designs. Fenced in areas ought to be intended to work with positive social connections among kangaroos.
   Gathering procedures consider the similarity, age, and social elements of people to advance a feeling of local area and lessen pressure.
   Noticing regular social progressive systems and considering social holding adds

to the prosperity of kangaroos. Very much planned territories incorporate elements that support social communications, for example, resting regions, taking care of stations, and adequate room for bunch exercises. Cautious observing of overall vibes permits guardians to mediate when important to guarantee the wellbeing and congruity of the kangaroo local area.

5. **Veterinary Consideration and Checking Offices**
   Compelling kangaroo territory configuration integrates offices for veterinary consideration and checking. Normal wellbeing checks, preventive consideration, and brief mediation in the event of wounds or diseases are basic to keeping up with the general strength of kangaroos in imprisonment.
   Assigned regions for veterinary assessments, quarantine, and clinical medicines are fundamental parts of kangaroo environments. Checking offices, including cameras and sensors, add to the continuous perception of kangaroo conduct, permitting guardians to recognize potential medical problems and change the executives rehearses likewise.
6. **Conceptive Contemplations and Rearing Offices**
   For foundations associated with hostage rearing projects, regenerative contemplations are essential in environment plan. Grasping the remarkable regenerative physiology of kangaroos, including early stage diapause, educates the creation regarding conditions helpful for rearing, birthing, and pocket advancement.
   Planning territories that oblige the conceptive necessities of kangaroos includes giving confined regions to birthing, pocket access, and secure spaces for raising posterity. Reproducing offices might incorporate separate nooks for viable matches, as well as regions intended to limit pressure during the rearing season.
7. **Instructive Open doors and Guest Experience**

Kangaroo environments in zoos and safe-havens frequently fill double needs, going about as instructive devices for guests. Very much planned environments consolidate highlights that permit guests to notice kangaroos participating in normal ways of behaving, advancing a more profound comprehension of these notorious marsupials.

Instructive signage, seeing stages, and interpretive showcases give data about kangaroo science, conduct, and preservation needs. Living space configuration ought to work with positive connections among guests and kangaroos while focusing on the prosperity and independence of the creatures.

3.3Behavioral enrichment and mental stimulation

Conduct improvement and mental excitement are urgent components under the watchful eye of creatures living in bondage, giving a way to upgrade their general prosperity. This approach recognizes the inborn requirements of creatures, planning to recreate normal ways of behaving and connect with their brains in a way that encourages actual wellbeing, smartness, and close to home satisfaction.

Figuring out Conduct Advancement

Characterizing Social Advancement:

Social improvement is a far reaching technique utilized under the watchful eye of hostage creatures, zeroing in on establishing conditions and exercises that support species-explicit ways of behaving. This training looks to ease the potential difficulties related with bondage, like restricted space and controlled conditions, by offering boosts that invigorate normal ways of behaving, mental commitment, and active work.

Significance of Regular Ways of behaving:

One of the essential objectives of social advancement is to give amazing open doors to creatures to communicate their regular ways of behaving. Hostage settings frequently force limitations on creatures' capacity to display instinctual exercises like scavenging, investigating, playing, and mingling. By presenting enhancement exercises, overseers expect to address these constraints, advancing a seriously satisfying and invigorating climate for the creatures.

Sorts of Social Advancement

Actual Improvement:

Actual advancement spins around adjusting the creatures' living spaces to energize development, exercise, and investigation. This can include the essential situation of climbing structures, rocks, trees, or adjustments in geography. By imitating normal scenes, actual advancement animates creatures to involve their bodies in manners that reflect their ways of behaving in the wild, advancing muscle improvement and generally wellness.

Mental Improvement:

Mental improvement focuses on the creatures' intellectual capacities, introducing difficulties that require critical thinking and navigation. Exercises like riddle feeders, stowed away treats, and novel items urge creatures to use their insight to acquire rewards. Mental enhancement is especially vital for species with elevated degrees of insight and complex mental capacities, cultivating mental feeling and commitment.

Social Enhancement:

Perceiving the social idea of numerous species, social enhancement intends to address the creatures' requirement for positive social communications. For normally friendly species, this includes setting out open doors for bunch exercises, giving friendship, and working with holding encounters. Social improvement adds to the general prosperity of creatures that depend on friendly designs for their psychological and close to home wellbeing.

Tactile Improvement:

Tactile improvement centers around invigorating the creatures' feelings of sight, smell, hearing, and contact. This can incorporate presenting new aromas, sounds, or visual boosts into their current circumstance. Tangible improvement gives assortment and oddity, forestalling tactile repetitiveness and offering a more powerful and connecting with experience for the creatures.

Conduct Enhancement in Various Settings

Zoos:

Present day zoos are progressively focusing on conduct enhancement as a basic piece of their creature care programs. With a double spotlight on preservation and training, zoos utilize different systems to improve the existences of their occupants.

Taking care of Enhancement: Zoos use puzzle feeders, disperse food, or conceal treats to support regular scrounging ways of behaving. This gives nourishment as well as draws in creatures in a way intelligent of their wild partners.

Primary Improvement: Giving climbing designs, ropes, and stages inside walled in areas copies the creatures' regular living spaces, offering valuable open doors for actual activity and investigation.

Mental Difficulties: Integrating toys, mirrors, or novel articles animates mental capacities and interest, giving mental commitment and forestalling weariness.

The mix of conduct improvement in zoos upgrades the prosperity of creatures, encourages regular ways of behaving, and fills in as an instructive device for guests, cultivating a more profound association with untamed life.

Untamed life Safe-havens:

Untamed life safe-havens, committed to giving asylum to saved or resigned creatures, focus on individual prosperity over open show. Conduct advancement in safe-havens is custom-made to address the extraordinary requirements and accounts of occupant creatures.

All encompassing Consideration: Asylums perceive the different foundations and social accounts of creatures, fitting advancement exercises appropriately. This approach recognizes the singularity of every inhabitant.

Non-Meddlesome Perception: Not at all like zoos, safe-havens take on a hands-off approach, permitting creatures to communicate regular ways of behaving without forcing pointless human connections. This advances a peaceful climate.

Species-Explicit Exercises: Improvement exercises are intended to take care of the particular requirements and ways of behaving of every species, recognizing their exceptional attributes and inclinations.

In safe-havens, the emphasis is on giving a solid and supporting climate where creatures can carry on with out their lives in conditions that look like their regular territories. Social advancement adds to the general prosperity of occupants, permitting them to communicate regular ways of behaving and find solace in their asylum homes.

Chapter 4

Nutritional Needs of Captive Kangaroos

Really focusing on kangaroos in imprisonment requires a profound comprehension of their dietary prerequisites to guarantee their wellbeing, prosperity, and generation. The interesting physiology, dietary propensities, and natural specialties of kangaroos present unmistakable difficulties and open doors for those liable for their consideration. This investigation digs into the nourishing necessities of hostage kangaroos, incorporating dietary structure, taking care of procedures, healthful difficulties, and the combination of exploration discoveries into reasonable administration.

1. **Life systems and Physiology**
1. **Dental Transformations**
   Kangaroos show dental transformations intelligent of their herbivorous eating regimen. Understanding their dental construction is pivotal for planning slims down that advance dental wellbeing and productive processing. The presence of consistently developing molars, proficient crushing capacities, and a specific perusing conduct all impact their wholesome necessities.
2. **Gastrointestinal Framework**

The kangaroo gastrointestinal framework is specific for the assimilation of stringy plant material. The foregut maturation process, special to macropods, makes light of a vital job in breaking cellulose. An assessment of the kangaroo's stomach compartments and microbial populaces gives experiences into their stomach related physiology and the ramifications for dietary preparation.

II. Dietary Arrangement

1. **Rummage Determination in Nature**
   Kangaroos, adjusted to the assorted biological systems of Australia, display particular rummaging ways of behaving. An examination of their normal eating routine, enveloping local grasses, peruse, and forbs, advises the improvement

regarding hostage abstains from food that reflect their wild partners. The occasional varieties in search accessibility and healthful substance additionally impact dietary contemplations.

2. **Large scale and Micronutrients**

Understanding the large scale and micronutrient necessities of hostage kangaroos is fundamental for figuring out adjusted consumes less calories. Looking at the jobs of proteins, starches, fats, nutrients, and minerals in kangaroo sustenance gives an establishment to making healthfully complete and species-explicit eating regimens.

III. Taking care of Procedures

1. **Peruse versus Field Diets**
   Hostage kangaroos might be given peruse, field, or a blend of both. Assessing the wholesome profiles of these feed sources and taking into account factors, for example, fiber content, protein quality, and energy thickness illuminates choices on the most proper taking care of methodology for various life stages and physiological states.
2. **Pelleted Diets and Enhancements**

Pelleted eats less and healthful enhancements assume a part in gathering the particular dietary necessities of hostage kangaroos. Analyzing the detailing and wholesome substance of financially accessible pellets and enhancements, and their reconciliation into hostage eats less, reveals insight into their adequacy in supporting wellbeing and generation.

IV. Regenerative Contemplations

1. **Lactation and Pocket Advancement**
   The regenerative physiology of female kangaroos presents one of a kind wholesome difficulties, especially during lactation and pocket improvement. Dissecting the expanded energy and supplement requests during these stages illuminates dietary changes in accordance with help effective generation in hostage settings.
2. **Rearing Projects and Hereditary Contemplations**

Hostage rearing projects add to the protection of kangaroo species. Assessing the nourishing parts of reproducing programs, including pre-rearing eating regimens, development, and the wholesome help for joeys, is fundamental. Hereditary contemplations affecting healthful necessities and it are likewise investigated to raise achievement.

V. Dietary Difficulties and Wellbeing Contemplations

1. **Weight and Metabolic Issues**
   In bondage, kangaroos might confront difficulties connected with stoutness and metabolic problems. Examining the variables adding to these issues, including diet piece, taking care of practices, and the effect on generally speaking well-being, directs the advancement of safeguard measures and mediation systems.
2. **Dental Wellbeing and Dietary Effect**

Dental wellbeing is vital to the general prosperity of hostage kangaroos. Investigating the connection between diet, dental wear, and oral wellbeing features the significance of giving proper surfaces and designs in the eating routine to advance dental life span.

VI. Incorporating Examination into Reasonable Administration

1. **Research Advances in Kangaroo Sustenance**
   The field of kangaroo sustenance persistently develops with progressing research. Analyzing late examinations on themes like supplement digestion, dietary inclinations, and conduct parts of taking care of illuminates proof based rehearses in hostage kangaroo the board.
2. **Functional Applications in Hostage Settings**

Making an interpretation of examination discoveries into functional applications is vital for overseers and administrators of hostage kangaroos. Examining the execution of examination driven dietary methodologies, taking care of conventions, and checking procedures improves the wellbeing and government assistance of kangaroos in imprisonment.

VII. Future Bearings and Preservation Suggestions

1. **Headways in Healthful Science**
   The fate of kangaroo nourishment in bondage includes headways in dietary science. Exploring arising research regions, for example, stomach microbiota organization, supplement bioavailability, and the utilization of novel feed sources, gives bits of knowledge into expected upgrades in hostage kangaroo counts calories.
2. **Protection Suggestions**

Perceiving the connection between sustenance, wellbeing, and proliferation in hostage settings has more extensive ramifications for the protection of kangaroo species. Pondering the job of all around fed and solid people in hostage populaces adds to the drawn out supportability of kangaroo protection endeavors.

4.1Dietary requirements for different kangaroo species

The kangaroo, a famous image of Australia, involves a few particular animal categories, each with remarkable qualities and dietary necessities. Understanding the dietary prerequisites of various kangaroo species is significant for their wellbeing, prosperity, and preservation. This far reaching investigation digs into the assorted dietary requirements of different kangaroo species, revealing insight into their transformative variations and the natural variables impacting their nourishing inclinations.

Transformative Foundation:

Kangaroos have developed north of millions of years, adjusting to Australia's assorted scenes. The different kangaroo species, like the red kangaroo (Macropus rufus), the eastern dim kangaroo (Macropus giganteus), and the western dim kangaroo (Macropus fuliginosus), have created particular dietary inclinations in view of their environments, taking care of ways of behaving, and developmental history. These transformations are fundamental for their endurance in conditions going from dry deserts to lavish fields.

Environment Impact on Dietary Inclinations:

One of the essential variables affecting the dietary necessities of kangaroo species is their living space. The red kangaroo, dominatingly tracked down in bone-dry districts, has adjusted to make due on meager vegetation and has created effective water-moderating systems. Contrastingly, the eastern dim kangaroo, possessing more mild and forested regions, has an eating routine wealthy in grasses and verdant vegetation. Understanding these living space explicit transformations is vital for laying out successful protection systems and hostage the board rehearses.

Nourishing Structure of Kangaroo Diets:

Kangaroos are herbivores with specific stomach related frameworks intended to separate greatest sustenance from plant material. Their eating routine principally comprises of grasses, forbs, and leaves. The dietary sythesis fluctuates across species, for certain kangaroos showing inclinations for specific plant types. The western dim kangaroo, for instance, is known to incline toward bushes and low-lying vegetation, while the red kangaroo is adjusted to consuming intense, sinewy grasses.

Dietary Varieties Among Kangaroo Species:

The red kangaroo, the biggest marsupial and the most notorious of its sort, fundamentally consumes dry, intense grasses. Its stomach related framework is enhanced to separate stringy plant material effectively, removing supplements from the bone-dry scenes it occupies. Then again, the eastern dark kangaroo is more versatile, consuming a more extensive scope of vegetation, including delicious grasses, leaves, and forbs. This adaptability permits the eastern dark kangaroo to flourish in various conditions, from forests to seaside regions.

Taking care of Ways of behaving and Searching Procedures:

Kangaroos show exceptional taking care of ways of behaving and rummaging procedures custom fitted to their individual surroundings. Touching, perusing, and particular taking care of are normal methodologies utilized by various species.

The red kangaroo, for example, depends on particular brushing, picking explicit plants with high dietary benefit. Conversely, the eastern dim kangaroo is known for its perusing conduct, using its prehensile lips to cull leaves and grasses from shrubs and trees. Understanding these taking care of ways of behaving is fundamental for keeping up with solid populaces in both normal and hostage settings.

Supplement Necessities and Metabolic Transformations:

Every kangaroo species has particular supplement prerequisites, directed by elements like age, sex, regenerative status, and ecological circumstances. Female kangaroos, particularly those with pocket youthful, have elevated nourishing requirements because of the energy requests of generation. Furthermore, metabolic variations, for example, the capacity to enter a condition of metabolic lethargy during food shortage, add to the kangaroo's wonderful capacity to get by in testing conditions.

Protection Suggestions:

Understanding the dietary prerequisites of various kangaroo species has huge ramifications for their preservation. Changes in land use, environment, and territory debasement can affect the accessibility of appropriate food sources. Protection endeavors should think about the insurance of natural surroundings as well as the safeguarding of the different plant species that structure the kangaroo's eating routine. Hostage the executives projects ought to repeat regular eating regimens as intently as could really be expected, considering the particular dietary inclinations of every species.

Hostage Kangaroo Diets:

In imprisonment, giving a healthfully adjusted diet that reflects the kangaroo's normal inclinations is vital for their wellbeing and prosperity. Zoos and untamed life safe-havens assume a urgent part in teaching people in general about kangaroo species while adding to explore on their dietary requirements. Hostage counts calories frequently incorporate a blend of high-fiber grasses, new peruse, and exceptionally formed kangaroo pellets to guarantee that the creatures get satisfactory sustenance.

Difficulties and Future Headings:

In spite of the continuous endeavors to comprehend and address the dietary necessities of various kangaroo species, challenges endure. Environmental change, territory misfortune, and human-natural life struggle present dangers to kangaroo populaces and their food sources. Continuous exploration is fundamental to adjust protection methodologies to the changing climate and guarantee the drawn out endurance of these notable marsupials.

4.2Challenges in providing a balanced diet

Guaranteeing a fair eating regimen is an essential part of advancing wellbeing and prosperity, yet various difficulties confound the quest for ideal sustenance. From financial requirements to social impacts, this investigation dives into the multi-layered difficulties that people and networks face in giving and keeping a decent eating routine. Understanding these difficulties is fundamental for creating compelling techniques to address sustenance inconsistencies and work on generally speaking general wellbeing.

Financial Boundaries:

One of the essential difficulties in giving a fair eating regimen is the monetary boundary that limits admittance to nutritious food sources. In numerous districts around the world, new natural products, vegetables, lean proteins, and entire grains can be more costly than handled and calorie-thick other options. This monetary dissimilarity causes what is going on where people with restricted monetary assets might turn to less nutritious, energy-thick food varieties, adding to the pervasiveness of diet-related medical problems like stoutness and unhealthiness.

Government arrangements and drives pointed toward financing good food sources, advancing nearby agribusiness, and guaranteeing reasonableness of nutritious choices can assist with easing monetary boundaries. Also, people group based programs that show financial plan agreeable dinner arranging and cooking abilities enable people to settle on better decisions inside their monetary requirements.

Social and Dietary Inclinations:

Social and dietary inclinations assume a huge part in forming people's dietary patterns. Conventional foods, family customs, and social practices can impact dietary decisions, once in a while impeding the reception of a decent eating regimen. Presenting wholesome training that regards social variety is critical in beating this test. By advancing the joining of solid options into conventional dishes and praising the social lavishness of different weight control plans, drives can overcome any issues between social inclinations and dietary requirements.

Speedy Ways of life:

Current ways of life, portrayed by furious timetables and time imperatives, represent a test to keeping a reasonable eating regimen. The comfort and accessibility of cheap food, pre-bundled dinners, and bites high in salt and sugar add to unfortunate dietary decisions. People confronting time imperatives might find it trying to focus on feast arranging, arrangement, and the utilization of supplement rich food sources.

Instructive missions underlining the significance of using time productively for dinner planning, alongside the advancement of speedy and sound recipes, can assist people with settling on better food decisions inside occupied plans.

Work environment drives that urge admittance to nutritious feasts and tidbits can likewise add to beating the difficulties related with quick moving ways of life.

Restricted Healthful Instruction:

Lacking wholesome schooling is an inescapable test that adds to unfortunate dietary propensities. Numerous people miss the mark on exhaustive comprehension of the healthful substance of food sources, segment control, and the drawn out ramifications of their dietary decisions. This information hole reaches out to perspectives, for example, perusing food marks, figuring out suggested day to day recompenses, and perceiving the effect of dietary examples on wellbeing.

Tending to this challenge requires a multi-layered approach, including the mix of nourishment instruction into school educational programs, local area based studios, and open web-based assets. Wellbeing experts, including dietitians and nutritionists, can assume a crucial part in scattering precise and viable data to enable people to settle on informed dietary decisions.

Food Deserts and Openness:

In different districts, the accessibility and availability of nutritious food sources present huge difficulties. Regions assigned as food abandons need admittance to supermarkets or markets that offer new produce and other solid choices. This restricted admittance is much of the time exacerbated by financial elements, further sustaining variations in sustenance.

Endeavors to address food deserts include local area based mediations, for example, laying out ranchers' business sectors, supporting nearby farming, and pushing for arrangements that boost the advancement of supermarkets in underserved regions. Versatile food drives and local area plants likewise add to expanding admittance to new and reasonable produce in regions where customary food retail foundation is deficient.

Misdirecting Promoting and Data:

The unavoidable impact of showcasing and deception in the food business represents a significant test to accomplishing a reasonable eating routine. Food items frequently go through broad showcasing, featuring their apparent medical advantages or dietary benefit. Nonetheless, a significant number of these cases can be misdirecting, driving buyers to go with decisions in view of wrong data.

Administrative measures to guarantee straightforward and precise food naming, combined with public mindfulness crusades that advance decisive pondering showcasing messages, are fundamental in tending to this test. Teaching buyers about how to observe solid dietary data engages them to settle on informed decisions, cultivating a culture of straightforwardness inside the food business.

Socio-Social and Natural Elements:

Socio-social and natural elements, including financial status, geographic area, and ecological maintainability, further confound the mission for a decent eating regimen. People confronting monetary difficulties might focus on quick reasonableness over long haul medical advantages. Furthermore, geographic varieties impact the accessibility of specific food sources, for certain districts being more helpful for the development of assorted crops.

Comprehensive methodologies that consider the interconnection of socio-social and natural variables are fundamental. Strategy drives advancing reasonable agribusiness, evenhanded admittance to assets, and addressing social determinants of wellbeing can add to conquering these difficulties. Coordinated effort between legislative bodies, non-administrative associations, and networks is essential for executing procedures that address the underlying drivers of dietary variations.

4.3Innovations in kangaroo nutrition

Kangaroos, notable images of Australia, have remarkable dietary necessities that assume a critical part in their wellbeing and prosperity. As the comprehension of kangaroo nourishment advances, imaginative methodologies are arising to address difficulties in hostage the executives, preservation, and generally kangaroo government assistance. This thorough investigation digs into the new developments in kangaroo nourishment, spreading over progressions in dietary details, taking care of procedures, and the joining of innovation to improve the strength of these striking marsupials.

1. **Transformative Variations and Dietary Prerequisites:**
   Prior to digging into ongoing developments, understanding the transformative variations and normal dietary prerequisites of kangaroos is fundamental. As herbivorous marsupials, kangaroos have particular stomach related frameworks intended to separate most extreme sustenance from stringy plant material. Their eating routine for the most part comprises of grasses, forbs, and leaves, with varieties across species in light of natural surroundings and ecological elements.
   The red kangaroo, for instance, occupies parched locales and has adjusted to consuming extreme, sinewy grasses, while the eastern dark kangaroo, tracked down in additional mild regions, has an eating regimen wealthy in delicious grasses, leaves, and forbs. These variations, created more than large number of years, feature the mind boggling connection among kangaroos and their common habitat.
2. **Challenges in Kangaroo Nourishment:**
   Regardless of their developmental transformations, kangaroos face wholesome difficulties, especially in hostage settings. Hostage kangaroo eats less carbs frequently battle to recreate the intricacy and assortment of the normal eating regimen, prompting worries about wholesome irregular characteristics, heftiness, and related medical problems. Also, the accessibility and nature of search in bondage may not generally meet the particular requirements of every kangaroo species.
   Hostage kangaroo populaces, whether in zoos, natural life safe-havens, or recovery focuses, require cautious dietary administration to guarantee the arrangement of a healthfully adjusted diet that upholds their physiological necessities. Developments in kangaroo nourishment expect to address these difficulties by consolidating logical examination, mechanical progressions, and a more profound comprehension of the healthful subtleties well defined for various kangaroo species.
3. **Ongoing Developments in Kangaroo Nourishment:**
1. **Definition of Custom-made Diets:**
   Ongoing developments in kangaroo nourishment include the definition of custom-made eats less that intently imitate the wholesome creation of their normal eating routine. This incorporates the distinguishing proof of key

supplements, like fiber, protein, nutrients, and minerals, fundamental for kangaroo wellbeing. Propels in nourishing science consider the advancement of particular kangaroo pellets or diets that take care of the remarkable prerequisites of various species.

These plans consider factors like age, sex, regenerative status, and generally wellbeing, giving a more redone way to deal with kangaroo nourishment. Zoos and untamed life safe-havens are progressively teaming up with nutritionists and specialists to make abstains from food that meet fundamental nourishing requirements as well as advance the general prosperity and regenerative outcome of hostage kangaroo populaces.

2. **Mix of Peruse and Advancement:**

   Emulating the scavenging conduct of wild kangaroos, creative methodologies remember the coordination of peruse and natural advancement for hostage settings. Peruse, which alludes to new, verdant branches and vegetation, gives physical and mental feeling to kangaroos. This approach expands their eating routine as well as advances normal ways of behaving, lessening pressure and upgrading their general government assistance.

   By consolidating peruse and improvement exercises, guardians mean to establish a more powerful and connecting with climate for hostage kangaroos. This supports their actual wellbeing as well as addresses social perspectives, adding to a more comprehensive way to deal with kangaroo prosperity in bondage.

3. **Use of Innovation in Dietary Checking:**

   Progressions in innovation assume a significant part in checking and enhancing kangaroo nourishment. Instruments, for example, wearable gadgets outfitted with sensors and cameras permit overseers and analysts to follow the movement levels, taking care of examples, and generally soundness of kangaroo populaces. These information driven experiences empower a more exact comprehension of individual dietary inclinations, metabolic rates, and wholesome necessities.

   Innovation additionally works with the advancement of brilliant taking care of frameworks that apportion food in view of individual kangaroo needs. Computerized taking care of stations, furnished with sensors and RFID (Radio-Recurrence Distinguishing proof) innovation, can perceive explicit kangaroos and administer redid segments, guaranteeing that every creature gets a suitable and adjusted diet.

4. **Dietary Exploration and Coordinated effort:**

Continuous wholesome exploration and cooperative endeavors between researchers, veterinarians, and overseers contribute essentially to developments in kangaroo sustenance. Concentrates on zeroed in on the stomach related physiology of kangaroos, supplement use, and the effect of various weight control plans on their wellbeing give significant experiences. This exploration shapes the reason for proof based dietary proposals and the advancement of nourishing rules for hostage kangaroo the executives.

Cooperative stages, like worldwide zoo affiliations, consider the sharing of best practices and encounters in kangaroo sustenance. These coordinated efforts encourage an aggregate way to deal with tending to normal difficulties and propelling the information base in kangaroo nourishment across various districts and establishments.

IV. Preservation Suggestions:

Developments in kangaroo nourishment have more extensive ramifications for the preservation of these marsupials, particularly with regards to protecting hereditary variety and guaranteeing the progress of rearing projects. Hostage rearing drives, frequently fundamental for species in danger, depend on ideal nourishment to help conceptive achievement and the soundness of posterity.

Furthermore, all around oversaw hostage populaces act as instructive diplomats, bringing issues to light about kangaroo protection and the significance of saving their normal territories. The information acquired from developments in kangaroo sustenance contributes not exclusively to the government assistance of hostage people yet in addition illuminates preservation methodologies pointed toward safeguarding kangaroo species in nature.

V. Difficulties and Future Bearings:

While developments in kangaroo sustenance show promising headways, challenges endure. The intricacy of imitating normal eating regimens in imprisonment, guaranteeing the supportability of peruse sources, and tending to the nourishing necessities of different kangaroo species stay continuous worries. Further exploration is expected to refine dietary definitions, improve taking care of techniques, and upgrade the general prosperity of kangaroos in imprisonment.

Future bearings in kangaroo nourishment might include investigating elective feed sources, examining the utilization of enhancements to address explicit dietary holes, and proceeding to refine innovation based checking and taking care of frameworks. Joint effort between establishments, progressing research drives, and a promise to executing proof based practices will be vital in defeating existing difficulties and driving proceeded with development in kangaroo sustenance.

Chapter 5

Veterinary Care and Health Management

Veterinary consideration and wellbeing the executives assume essential parts in guaranteeing the prosperity of creatures across assorted species. As how we might interpret creature science, conduct, and medical services propels, so does the requirement for complete veterinary consideration. This broad investigation dives into different parts of veterinary consideration and wellbeing the executives, enveloping preventive measures, analytic procedures, treatment modalities, and the advancing job of veterinarians in advancing creature government assistance.

1. **The Job of Veterinarians:**
1. **Verifiable Point of view:**
   The act of veterinary medication goes back hundreds of years, with early veterinarians principally centered around treating domesticated animals and working creatures. Over the long haul, the calling has developed, extending to cover a great many creatures, including friend creatures, fascinating species, and untamed life. Present day veterinarians go through thorough preparation, furnishing them with the abilities to address assorted medical problems and add to progressions in veterinary science.
2. **Veterinary Specializations:**
   As the field of veterinary medication has developed, specific branches have arisen to address the remarkable necessities of various animal species. These claims to fame incorporate little creature medication, equine medication, intriguing creature medication, natural life medication, and that's only the tip of the iceberg. Every specialization requires explicit information and aptitude, mirroring the broadness and profundity of veterinary consideration.
3. **The One Wellbeing Approach:**

The idea of "One Wellbeing" highlights the interconnectedness of human, creature, and ecological wellbeing. Veterinarians assume an essential part in this interdisciplinary methodology, perceiving that the soundness of people and creatures is personally connected. By tending to zoonotic illnesses, advancing capable anti-infection use, and adding to worldwide wellbeing drives, veterinarians add to an all encompassing comprehension of wellbeing that reaches out past individual species.

II. Preventive Veterinary Consideration:

1. **Inoculation Projects:**
   Preventive veterinary consideration frequently starts with inoculation programs pointed toward shielding creatures from irresistible sicknesses. Immunizations have been instrumental in controlling and annihilating sicknesses like rabies, sickness, and parvovirus in homegrown creatures. Also, natural life preservation endeavors frequently include immunization projects to protect jeopardized species from infection flare-ups.
2. **Parasite Control:**
   Parasite invasions represent a critical danger to the strength of creatures. Veterinarians utilize different preventive measures, including deworming drugs, insect and tick control, and methodologies to moderate the effect of inward and outside parasites. Normal screenings and altered parasite control plans are fundamental parts of preventive veterinary consideration.
3. **Sustenance and Dietary Advising:**
   Legitimate sustenance is basic to creature wellbeing. Veterinarians give dietary guiding customized to the particular requirements of every species, age gathering, and individual medical issue. From planning adjusted diets to tending to nourishing lacks, veterinarians assume a basic part in advancing ideal wellbeing through dietary administration.
4. **Dental Consideration:**

Dental wellbeing is frequently ignored yet is basic to the general prosperity of creatures. Veterinarians direct dental assessments, cleanings, and medicines to resolve dental issues like periodontal infection. Dental consideration is particularly urgent for friend creatures, as dental issues can prompt foundational medical problems.

III. Analytic Procedures in Veterinary Medication:

1. **Imaging Modalities:**
   Headways in imaging advancements have changed analytic capacities in veterinary medication. X-beams, ultrasound, CT sweeps, and X-ray are utilized to envision inner designs, recognize anomalies, and guide careful mediations. These harmless strategies assume a significant part in diagnosing conditions going from cracks to cancers.
2. **Lab Diagnostics:**
   Research center diagnostics incorporate a great many tests, including bloodwork,

urinalysis, and microbiological investigations. These tests give significant bits of knowledge into the general strength of creatures, help in the analysis of illnesses, and assist with observing treatment reactions. Veterinarians depend on research center outcomes to arrive at informed conclusions about persistent consideration.

3. **Sub-atomic Diagnostics:**
   The field of sub-atomic diagnostics has extended in veterinary medication, considering the location of hereditary problems, irresistible specialists, and other atomic markers. Polymerase chain response (PCR) and DNA sequencing are among the methods used to dissect hereditary material, empowering veterinarians to make exact judgments and execute designated treatment plans.
4. **Endoscopy and Negligibly Obtrusive Systems:**

Endoscopy and negligibly obtrusive systems have become norm in veterinary medication for diagnosing and treating different circumstances. Endoscopes are utilized to imagine inside organs, get biopsies, and carry out negligibly intrusive procedures. These methods diminish recuperation times and improve symptomatic precision.

IV. Treatment Modalities in Veterinary Medication:

1. **Pharmacotherapy:**
   Pharmacotherapy is a foundation of veterinary medication, including the utilization of prescriptions to treat a great many circumstances. Veterinarians endorse anti-microbials, mitigating drugs, pain killers, and different meds in light of the particular requirements of every patient. Dependable and sensible utilization of drugs is critical to stay away from opposition and antagonistic impacts.
2. **Medical procedure:**
   Careful mediations are in many cases important to address wounds, intrinsic irregularities, and certain ailments. Progresses in careful methods, sedation conventions, and postoperative consideration add to further developed results for careful patients. Veterinarians might perform delicate tissue medical procedures, muscular strategies, and insignificantly intrusive medical procedures relying upon the case.
3. **Restoration and Active recuperation:**
   Restoration and active recuperation have earned respect in veterinary medication as necessary parts of treatment plans. These modalities assist creatures with recuperating from wounds, medical procedures, or persistent circumstances.
   Actual advisors work close by veterinarians to plan recovery programs that further develop portability, diminish agony, and improve in general personal satisfaction.
4. **Option and Integral Treatments:**

Notwithstanding ordinary medicines, veterinarians might integrate option and reciprocal treatments to address specific circumstances. Needle therapy, chiropractic care, home grown medication, and physiotherapy are instances of modalities that supplement conventional veterinary methodologies. Incorporating these treatments into complete treatment plans considers a comprehensive way to deal with creature wellbeing.

V. Geriatric and Palliative Consideration:

As creatures age, they might encounter age-related medical problems that require particular consideration. Geriatric veterinary medication centers around overseeing conditions like joint pain, mental brokenness, and organ brokenness in senior creatures. Palliative consideration plans to work on the personal satisfaction for creatures with persistent or terminal sicknesses, stressing torment the board and steady consideration.

VI. Crisis and Basic Consideration:

Crisis and basic consideration are fundamental parts of veterinary medication, tending to pressing clinical circumstances and giving life-saving intercessions. Crisis veterinarians are prepared to deal with injury cases, poison levels, and intense clinical emergencies. Basic consideration offices furnished with cutting edge observing and life emotionally supportive networks assume a significant part in overseeing fundamentally sick patients.

VII. The Human-Creature Bond:

The human-creature bond is a focal topic in veterinary consideration, perceiving the profound and mental associations among people and their creature buddies. Veterinarians assume a steady part in assisting pet people with exploring end-of-life choices, adapt to misfortune, and give ideal consideration to their creature mates. This part of veterinary medication highlights the significance of sympathy, correspondence, and grasping in the veterinarian-client relationship.

VIII. Untamed life and Protection Medication:

Veterinary consideration reaches out past homegrown creatures to envelop natural life and protection endeavors. Natural life veterinarians add to the wellbeing and preservation of imperiled species, leading exploration, tending to illness flare-ups, and executing methodologies to safeguard biodiversity. Protection medication accentuates the interconnectedness of biological system wellbeing, creature wellbeing, and human wellbeing.

IX. Challenges in Veterinary Consideration and Wellbeing The executives:

1. **Zoonotic Illnesses:**
   The development and transmission of zoonotic illnesses present continuous difficulties in veterinary consideration. Illnesses that can be communicated among creatures and people, like avian flu and West Nile infection, require cautious

reconnaissance, preventive measures, and cooperative endeavors among veterinary and general wellbeing experts.

2. **Anti-infection Obstruction:**
   Anti-infection obstruction is a worldwide worry in both human and veterinary medication. The dependable utilization of anti-infection agents is vital to forestall the advancement of safe types of microscopic organisms. Veterinarians assume a key part in advancing antimicrobial stewardship, upholding for reasonable anti-infection use, and investigating elective procedures for overseeing bacterial diseases.
3. **Moral Situations and End-of-Life Care:**
   Veterinarians frequently face moral situations connected with treatment choices, asset assignment, and end-of-life care. Speaking with pet people about treatment choices, anticipation, and personal satisfaction contemplations requires awareness and empathy. Moral conversations encompassing willful extermination and high level consideration arranging are indispensable parts of veterinary practice.
4. **Monetary Requirements:**
   Monetary requirements can restrict admittance to veterinary consideration for a few pet people, prompting difficulties in giving ideal medical care to their creatures. Veterinarians should explore these requirements while endeavoring to convey the most ideal consideration inside the accessible assets. Local area outreach projects and drives that advance monetary help can assist with tending to this test.

   D. Emotional wellness and Prosperity of Veterinarians:
   The requesting idea of veterinary practice, combined with personal difficulties connected with patient results and client communications, can affect the emotional wellness and prosperity of veterinarians. Perceiving the significance of psychological wellness support, proficient associations and establishments are progressively underlining assets and projects to address these worries.
5. **Quick Headways in Veterinary Innovation:**

While mechanical progressions upgrade demonstrative and treatment abilities, the fast speed of advancement presents difficulties in keeping veterinarians side by side of new turns of events.

Proceeding with instruction, preparing programs, and cooperative organizations are fundamental to guarantee that veterinary experts can use the advantages of arising innovations actually.

X. Future Headings and Developments in Veterinary Consideration:

1. **Telemedicine:**
   Telemedicine is acquiring unmistakable quality in veterinary consideration, giving open doors to distant discussions, follow-up arrangements, and client

training. This innovation improves openness to veterinary mastery, especially in underserved regions. Be that as it may, contemplations connected with information security, administrative structures, and the impediments of virtual assessments remain subjects of progressing conversation.

2. **Customized Medication:**
   Headways in hereditary examination and sub-atomic diagnostics are making ready for customized medication in veterinary consideration. Fitting treatment plans in view of a singular creature's hereditary cosmetics considers designated mediations and more exact administration of sicknesses. Customized medication holds guarantee for advancing treatment results and limiting unfavorable impacts.
3. **Man-made brainpower and Information Investigation:**
   The combination of computerized reasoning (man-made intelligence) and information examination into veterinary practice improves demonstrative exactness, treatment arranging, and wellbeing checking. Artificial intelligence calculations can investigate clinical pictures, decipher research center outcomes, and give choice help to veterinarians. These innovations add to more proficient and information driven ways to deal with veterinary consideration.
4. **Regenerative Medication:**
   Regenerative medication, including foundational microorganism treatment and tissue designing, is arising as a promising field in veterinary consideration. These methodologies expect to bridle the regenerative capability of cells and tissues to treat wounds and degenerative circumstances. Regenerative medication holds potential for further developing results in muscular cases, wound recuperating, and different circumstances.
5. **Integrative and All encompassing Methodologies:**

Integrative and all encompassing ways to deal with veterinary consideration keep on acquiring prevalence. Consolidating customary medicines with elective treatments like needle therapy, natural medication, and exercise based recuperation offers a more far reaching and patient-focused approach. Integrative veterinary consideration thinks about the singular requirements of every patient, advancing in general prosperity.

5.1Common health issues in captive kangaroos

Hostage kangaroos, housed in zoos, natural life safe-havens, and recovery focuses, face a novel arrangement of wellbeing challenges particular from their partners in nature. The progress from normal territories to hostage conditions presents different elements that can affect the physical and mental prosperity of kangaroos. This investigation dives into normal medical problems looked by hostage kangaroos, including perspectives like sustenance, regenerative wellbeing, social worries, and irresistible sicknesses. Understanding these difficulties is pivotal for creating viable administration techniques to guarantee the wellbeing and government assistance of hostage kangaroo populaces.

1. **Healthful Difficulties:**
1. **Diet Abberations:**
   One of the essential wellbeing worries for hostage kangaroos rotates around diet inconsistencies. The normal eating regimen of kangaroos in the wild comprises of a different scope of grasses, forbs, and leaves, giving fundamental supplements and strands. In imprisonment, recreating this differed diet can be testing, prompting possible wholesome awkward nature. Deficient admittance to peruse and an absence of variety in scrounge choices might bring about lacks in nutrients, minerals, and other crucial supplements.
2. **Dental Issues:**
   Dental issues can emerge in hostage kangaroos because of diet-related factors. The restricted assortment of food things and the shortfall of rough scrounge can add to dental issues like malocclusion, congested teeth, and dental illness. These issues might influence the kangaroo's capacity to bite food appropriately, prompting wholesome inadequacies and influencing by and large wellbeing.
3. **Stoutness:**

Insufficient eating routine control and restricted open doors for regular scrounging can incline hostage kangaroos toward weight. Heftiness is a huge wellbeing worry, as it can prompt a scope of optional issues, including joint issues, cardiovascular issues, and diminished regenerative achievement.

Keeping a decent eating routine and carrying out natural enhancement exercises that energize active work are fundamental in forestalling weight in hostage kangaroos.

II. Conceptive Wellbeing:

1. **Stress-Initiated Conceptive Issues:**
   Hostage conditions can prompt pressure in kangaroos, affecting their regenerative wellbeing. Stress might prompt disturbances in the regenerative cycle, unpredictable estrous cycles, and, surprisingly, conceptive concealment. Female kangaroos might encounter trouble imagining, prompting decreased rearing achievement. Tending to stressors through appropriate ecological plan, social designs, and conduct advancement is essential to help regenerative wellbeing.
2. **Barrenness:**
   Barrenness can be a worry in hostage kangaroo populaces, and it might result from different elements, including wholesome lopsided characteristics, stress, age-related issues, and hidden ailments. Checking conceptive wellbeing, leading ripeness appraisals, and carrying out designated mediations, like dietary changes

and stress decrease procedures, are fundamental in overseeing and forestalling fruitlessness in hostage kangaroos.

3. **Conceptive Lot Problems:**

Female kangaroos might be powerless to conceptive parcel problems, like blisters or uterine diseases. These circumstances can influence fruitfulness and may prompt inconveniences during pregnancy. Standard veterinary assessments, including regenerative wellbeing evaluations and ultrasound checking, are crucial for early location and the board of conceptive parcel issues in hostage female kangaroos.

III. Conduct Concerns:

1. **Stereotypic Ways of behaving:**
   Hostage kangaroos might show stereotypic ways of behaving, tedious activities that fill no obvious need and are frequently connected with pressure or disappointment. Normal stereotypic ways of behaving in kangaroos incorporate pacing, head swaying, and self-prepping. These ways of behaving might be demonstrative of natural stressors, lacking space, or insufficient social collaborations. Carrying out natural advancement and guaranteeing appropriate social designs can assist with limiting stereotypic ways of behaving in hostage kangaroos.
2. **Social Pressure:**
   Kangaroos are social creatures in the wild, and their prosperity is intently attached to social collaborations. In imprisonment, deficient social designs, stuffing, or the shortfall of proper friendship can prompt social pressure. This pressure might appear in forceful ways of behaving, withdrawal, or the advancement of stereotypic ways of behaving. Appropriately planned nooks, social groupings, and cautious perception of individual kangaroo communications are fundamental for tending to social pressure.
3. **Natural Improvement:**

Giving adequate natural enhancement is basic for advancing mental excitement and forestalling social issues in hostage kangaroos. Enhancement exercises might incorporate the presentation of normal substrates, the arrangement of puzzle feeders, and the joining of climbing structures. These exercises lighten weariness as well as draw in the kangaroo's regular impulses, adding to generally conduct prosperity.

IV. Irresistible Illnesses:

1. **Parasitic Diseases:**
   Parasitic diseases are normal wellbeing worries in hostage kangaroos. Inside parasites, like gastrointestinal worms, can represent a danger to the soundness of kangaroos, prompting weight reduction, shortcoming, and stomach related

issues. Standard waste assessments, vital deworming programs, and keeping up with clean nooks are significant in forestalling and overseeing parasitic contaminations.

2. **Respiratory Contaminations:**
   Hostage kangaroos might be powerless to respiratory contaminations, particularly in conditions with poor ventilation or openness to ecological pollutants. Respiratory contaminations can prompt side effects like hacking, nasal release, and dormancy. Keeping up with clean nooks, giving satisfactory ventilation, and speedily tending to any indications of respiratory misery are fundamental in forestalling and dealing with these contaminations.
3. **Anticipation and Biosecurity:**

Carrying out severe biosecurity measures is urgent in forestalling the presentation and spread of irresistible sicknesses in hostage kangaroo populaces. This incorporates quarantine conventions for fresh introductions, normal wellbeing screenings, and observing for any indications of sickness. Legitimate cleanliness rehearses, like sterilization of hardware and nooks, add to keeping a solid and sickness free climate.

V. Foot and Paw Issues:

1. **Pododermatitis:**
   Pododermatitis, or irritation of the footpads, can be a worry in hostage kangaroos, especially in those housed on hard surfaces. This condition might prompt inconvenience, weakness, and open injuries. Furnishing substrates with suitable padding, standard assessment of foot wellbeing, and addressing natural factors that add to pododermatitis are fundamental preventive measures.
2. **Nail and Hook Issues:**

Kangaroos utilize their strong rear appendages for bouncing and development, and issues with nails or paws can emerge in bondage. Congested nails, wounds, or diseases might happen, affecting the kangaroo's capacity to easily move. Routine nail manages, normal foot examinations, and keeping up with appropriate substrates can help forestall and address nail and hook issues in hostage kangaroos.

VI. Veterinary Consideration and Observing:

1. **Routine Wellbeing Assessments:**
   Routine wellbeing assessments led by qualified veterinarians are vital for observing the general prosperity of hostage kangaroos. These assessments incorporate actual appraisals, dental checks, regenerative wellbeing evaluations, and screenings for normal medical problems. Early recognition of wellbeing concerns considers convenient intercession and preventive measures.

2. **Individualized Care Plans:**
   Every hostage kangaroo might have special wellbeing necessities in view of variables like age, sex, conceptive status, and in general wellbeing. Creating individualized care designs that address explicit wholesome necessities, natural advancement inclinations, and wellbeing contemplations is fundamental for advancing the prosperity of every kangaroo in bondage.
3. **Cooperative Methodologies:**

Coordinated effort between veterinarians, creature care staff, and untamed life specialists is fundamental in tending to the different wellbeing challenges looked by hostage kangaroos. Ordinary correspondence, sharing of perceptions, and cooperative critical thinking add to a thorough and successful way to deal with veterinary consideration and wellbeing the executives.

VII. Schooling and Public Mindfulness:

Teaching overseers, animal specialists, and general society about the particular requirements and difficulties of hostage kangaroos is significant for advancing mindful consideration and preservation endeavors. Public mindfulness drives can assist with gathering support for preservation programs, subsidizing for veterinary consideration, and the execution of best practices in hostage kangaroo the board.

5.2Preventive healthcare measures

Preventive medical care is a proactive and comprehensive methodology pointed toward keeping up with and advancing generally prosperity by forestalling the beginning of infections and recognizing potential wellbeing gambles from the get-go. These actions are intended to decrease the weight of sickness, work on the personal satisfaction, and add to longer, better lives. This investigation dives into the meaning of preventive medical services, key parts, and the positive effect it can have on people and networks.

1. **Meaning of Preventive Medical care:**
1. **Early Identification and Intercession:**
   Preventive medical care underscores early location and mediation, permitting medical services suppliers to distinguish potential medical problems before they become more extreme or constant. Normal screenings, wellbeing check-ups, and vaccinations empower medical care experts to evaluate risk factors, distinguish irregularities, and mediate immediately, frequently forestalling the movement of illnesses.
2. **Cost-Proficiency:**
   Preventive medical services measures are many times more practical than treating progressed sicknesses. By putting resources into preventive procedures, people and medical care frameworks can lessen the monetary weight related with the therapy of preventable circumstances. Normal check-ups, immunizations, and

way of life intercessions can forestall the requirement for expensive operations and long haul care.

3. **Worked on Personal satisfaction:**

Focusing on preventive medical services adds to a better personal satisfaction. By embracing sound ways of behaving, overseeing risk factors, and tending to wellbeing concerns early, people can keep up with their physical and mental prosperity. An emphasis on preventive measures permits individuals to take part in dynamic and satisfying ways of life, cultivating a feeling of essentialness and strength.

II. Parts of Preventive Medical services:

1. **Immunizations:**
   Vaccinations are essential preventive estimates that shield people from irresistible infections. Immunizations invigorate the invulnerable framework to deliver a resistant reaction, giving insusceptibility without causing the actual illness. Routine inoculations, as suggested by medical care specialists, are essential for people, everything being equal, adding to local area invulnerability and forestalling the spread of infectious sicknesses.
2. **Ordinary Wellbeing Check-ups:**
   Ordinary wellbeing check-ups are central in preventive medical services. These assessments, led by medical services experts, incorporate evaluations of indispensable signs, screenings for normal medical problems, and conversations about way of life factors. Normal check-ups give a chance to recognize risk factors, screen generally speaking wellbeing, and lay out a pattern for future correlations.
3. **Screenings and Demonstrative Tests:**
   Screenings and analytic tests are explicit assessments led to recognize early indications of sicknesses or conditions. Models incorporate mammograms for bosom malignant growth, colonoscopies for colorectal disease, and blood tests for diabetes. These screenings are customized to individual gamble elements, age, and orientation, taking into consideration designated anticipation and early mediation.
4. **Way of life Changes:**
   Solid way of life decisions are indispensable to preventive medical services. Empowering people to take on ways of behaving like customary activity, a reasonable eating regimen, satisfactory rest, and stress the board can essentially influence by and large wellbeing. Way of life changes address risk factors related with constant circumstances like coronary illness, diabetes, and corpulence.
5. **Wellbeing Training and Advancement:**

Wellbeing schooling assumes a vital part in preventive medical care by expanding mindfulness and advancing informed direction. Instructive drives give people the information and apparatuses to settle on better decisions, oversee pressure, and figure out the significance of preventive measures. General wellbeing efforts and local area outreach add to far and wide wellbeing advancement.

III. Preventive Medical services Across the Life expectancy:

1. **Adolescence:**
   Preventive medical care starts in youth, zeroing in on immunizations, routine pediatric check-ups, and screenings for formative achievements. Early mediations can resolve issues like unhealthiness, youth corpulence, and learning handicaps. Laying out sound propensities in youth establishes the groundwork for a long period of prosperity.
2. **Pre-adulthood:**
   Youths benefit from preventive medical care estimates that address the interesting difficulties of this life stage. Vaccinations, sexual wellbeing training, and screenings for emotional well-being concerns become significant parts. Empowering sound way of life decisions during immaturity can lay out sure propensities that persevere into adulthood.
3. **Adulthood:**
   Preventive medical care for grown-ups includes normal wellbeing check-ups, screenings for constant circumstances, and age-proper inoculations. Way of life changes, including keeping a sound weight, overseeing pressure, and keeping away from tobacco and inordinate liquor utilization, add to generally speaking prosperity. Screenings for conditions like hypertension, diabetes, and certain tumors become more basic with age.
4. **Seniors:**

In more established grown-ups, preventive medical care shifts concentration to address age-related wellbeing concerns. Standard screenings for conditions like osteoporosis, vision and hearing debilitations, and mental degradation become vital. Vaccinations for flu and pneumonia are prescribed to safeguard seniors from respiratory contaminations, which can be especially serious in this age bunch.

IV. Key Difficulties in Executing Preventive Medical services:

1. **Admittance to Medical services:**
   Admittance to preventive medical services stays a test in different locales worldwide. Restricted admittance to medical services offices, monetary imperatives, and geographic boundaries can obstruct people from getting opportune screenings and check-ups. Tending to medical care inconsistencies and further

developing availability are fundamental parts of complete preventive medical care methodologies.

2. **Wellbeing Proficiency:**
   Wellbeing education, the capacity to comprehend and apply wellbeing data, assumes a critical part in preventive medical services. Restricted wellbeing proficiency can ruin people from settling on informed conclusions about their wellbeing, grasping clinical directions, and perceiving the significance of preventive measures. Advancing wellbeing training and clear correspondence in medical services settings is pivotal for beating this test.
3. **Adherence to Preventive Measures:**
   Empowering people to stick to preventive measures, for example, immunization plans, customary screenings, and way of life alterations, can challenge. Factors like social convictions, confusions, and individual inclinations might impact adherence. Executing compelling correspondence systems, tending to social awarenesses, and cultivating patient commitment add to expanded adherence.
4. **Conduct Change:**

Starting and supporting conduct change is an intricate cycle. Numerous preventive medical services measures include way of life adjustments, which expect people to embrace new propensities and schedules. Conduct change intercessions, including persuasive advising, support gatherings, and customized objective setting, are fundamental parts of preventive medical care procedures.

V. Future Bearings in Preventive Medical care:

1. **Mechanical Advancements:**
   Headways in innovation, including wearable gadgets, telemedicine, and wellbeing checking applications, are changing the scene of preventive medical care. These advancements give people instruments for self-observing, customized wellbeing information, and distant discussions, improving the openness and adequacy of preventive measures.
2. **Hereditary and Accuracy Medication:**
   The mix of hereditary data into preventive medical services considers more customized and exact intercessions. Hereditary testing can recognize people at higher gamble for specific circumstances, empowering designated preventive measures and early intercessions. Accuracy medication tailors medical services procedures in light of a singular's exceptional hereditary cosmetics, upgrading the adequacy of preventive methodologies.
3. **Local area Based Intercessions:**

Local area based preventive medical services mediations include connecting with nearby networks in wellbeing advancement drives. These drives might incorporate wellbeing programs, instructive studios, and cooperative endeavors between medical services suppliers and local area associations. Local area contribution encourages a steady climate for preventive measures and addresses social determinants of wellbeing.

5.3Role of veterinarians and caretakers

The job of veterinarians and guardians in the prosperity of creatures, whether in homegrown settings, zoos, natural life asylums, or recovery focuses, is multi-layered and basic. These committed experts assume fundamental parts in guaranteeing the wellbeing, government assistance, and generally personal satisfaction for creatures under their consideration.

Veterinarians, with their specific preparation in creature medication, are answerable for diagnosing and treating sicknesses, directing routine wellbeing assessments, and carrying out surgeries when essential. Their skill reaches out to preventive consideration, where they regulate inoculations, carry out parasite control gauges, and foster customized nourishment plans. Notwithstanding clinical mediations, veterinarians add to the administration of creature populaces through conceptive wellbeing projects and aid the protection of imperiled species.

Guardians, frequently working intimately with veterinarians, are answerable for the everyday consideration and prosperity of creatures. They guarantee that creatures get legitimate sustenance, spotless and safe living conditions, and open doors for mental and actual feeling. Guardians assume an essential part in noticing and observing creature conduct, immediately distinguishing indications of trouble or disease. Their collaborations with creatures add to the socialization and enhancement of hostage populaces, cultivating a feeling of trust and prosperity.

Coordinated effort among veterinarians and guardians is fundamental for extensive creature care. Standard correspondence, preparing projects, and joint direction add to the general wellbeing and government assistance of the creatures. What's more, instruction and effort endeavors by the two veterinarians and overseers assist with bringing issues to light about capable pet proprietorship, untamed life protection, and the significance of moral and altruistic treatment of creatures.

Eventually, the joined endeavors of veterinarians and overseers establish a climate that advances ideal wellbeing, forestalls infection, and guarantees the prosperity of creatures under human consideration. Their commitment and skill are essential in keeping a harmony between protection, training, and moral treatment of creatures in different settings.

Chapter 6

Reproduction and Breeding Programs

Proliferation and rearing projects assume a urgent part in the administration and preservation of creature populaces across different species. This exhaustive investigation dives into the mind boggling universe of generation, enveloping normal cycles, helped regenerative advancements, and the foundation of reproducing programs. From the moves looked by hostage rearing drives to the moral contemplations encompassing hereditary variety and government assistance, this assessment expects to reveal insight into the complex parts of multiplication and reproducing in both homegrown and wild settings.

1. **Regular Propagation Cycles:**
1. **Mating Conduct:**
   In the wild, mating conduct is affected by different elements, including hormonal changes, ecological signals, and social elements. Understanding the normal mating ways of behaving of various species is fundamental for fruitful propagation. Romance customs, mate determination, and mating systems fluctuate broadly among creatures, reflecting variations to their biological specialties.
2. **Estrus Cycles and Regenerative Chemicals:**
   Numerous species display explicit regenerative cycles managed by hormonal vacillations. Estrus cycles, portrayed by changes in chemical levels, impact the receptivity of females to mating. Understanding these cycles is vital for deciding ideal rearing times. Conceptive chemicals, for example, estrogen and progesterone, assume key parts in organizing the physiological changes related with pregnancy and parturition.
3. **Pregnancy and Parturition:**
   The development time frame and parturition processes shift altogether among species. From the generally short incubation of certain rodents to the lengthy pregnancies of elephants, every species has developed conceptive procedures fit

to its natural specialty. Pregnancy and parturition include complex cooperations between maternal physiology, fetal turn of events, and natural variables.

4. **Parental Consideration:**

Parental consideration is a basic part of normal propagation, guaranteeing the endurance and prosperity of posterity. Various species display assorted nurturing ways of behaving, going from the concentrated consideration given by warm blooded animals to the shifted procedures utilized by birds, reptiles, and bugs. Parental speculation and providing care ways of behaving add to the regenerative progress of people and populaces.

II. Helped Regenerative Innovations (Craftsmanship):

1. **In Vitro Preparation (IVF):**
   In hostage rearing projects and helped propagation, In Vitro Preparation (IVF) is a broadly utilized method. It includes the assortment of eggs and sperm, which are then joined in a research center setting. The subsequent undeveloped organisms are refined prior to being embedded into a proxy or the natural mother. IVF has been instrumental in the protection of imperiled species with regenerative difficulties.
2. **Planned impregnation (man-made intelligence):**
   Planned impregnation (man-made intelligence) is a procedure where sperm is gathered and brought into the regenerative parcel of a female without regular mating. Computer based intelligence is utilized in different settings, including domesticated animals rearing, protection programs, and the administration of homegrown creature populaces. It takes into account the controlled appropriation of hereditary material and works with rearing without the requirement for direct actual contact.
3. **Undeveloped organism Move:**
   Undeveloped organism move includes the assortment and control of undeveloped organisms, trailed by their exchange to a substitute mother. This strategy is important for expanding conceptive rates, especially in circumstances where normal reproducing might challenge. Undeveloped organism move is used in animals rearing, jeopardized species protection, and improving the conceptive productivity of important hereditary lines.
4. **Cryopreservation of Gametes and Undeveloped organisms:**

Cryopreservation is the method involved with safeguarding gametes (sperm and eggs) or incipient organisms at super low temperatures. This method considers long haul stockpiling and works with the making of hereditary banks. Cryopreservation is fundamental for keeping up with hereditary variety, overseeing reproducing programs, and preserving the hereditary material of imperiled species.

III. Hostage Reproducing Projects:

1. **Protection Objectives:**
   Hostage rearing projects are laid out with the essential objective of moderating species confronting dangers in nature. These dangers might incorporate territory misfortune, poaching, sickness, and environmental change. By keeping up with practical populaces in bondage, these projects act as protection against termination and add to the possible renewed introduction of people into their regular natural surroundings.
2. **Hereditary Variety:**
   Keeping up with hereditary variety is a focal thought in hostage reproducing programs. Hereditary variety guarantees the versatility and strength of populaces to changing ecological circumstances. Program supervisors utilize systems like family investigation, hereditary checking, and the foundation of studbooks to go with informed reproducing choices and keep away from inbreeding discouragement.
3. **Populace The board:**
   Viable populace the executives is essential for the progress of hostage rearing projects. Administrators should consider factors, for example, sex proportions, age structures, and regenerative wellness while arranging rearing matches. Executing techniques to limit pressure, give appropriate living spaces, and elevate regular ways of behaving adds to the general prosperity of creatures in imprisonment.
4. **Renewed introduction and Delivery:**

A definitive objective of numerous hostage reproducing programs is the fruitful renewed introduction of people into their local territories. Renewed introduction endeavors include cautious preparation, living space reclamation, and observing of delivered people. Achievement is estimated by the quantity of people delivered as well as by their capacity to flourish and add to wild populaces.

IV. Challenges in Multiplication and Rearing Projects:

1. **Conduct Difficulties:**
   The progress from normal territories to imprisonment can present conduct difficulties for creatures in rearing projects. Species-explicit ways of behaving connected with mating, settling, and raising posterity might be impacted by the limitations of imprisonment. Tending to these difficulties requires figuring out the regular ways of behaving of the species and giving fitting natural improvement.
2. **Conceptive Medical problems:**
   Hostage conditions might current circumstances that influence conceptive

wellbeing, including pressure, dietary awkward nature, and restricted space for proactive tasks. These variables can prompt conceptive anomalies, diminished fruitfulness, and difficulties in effective reproducing. Preventive veterinary consideration, sustenance the board, and social advancement are fundamental parts of tending to regenerative medical problems.

3. **Inbreeding Despondency:**
   Keeping up with hereditary variety in hostage populaces is critical to forestalling inbreeding gloom. Inbreeding, or mating between firmly related people, can prompt a decrease in wellness and an expanded weakness to illnesses. Hereditary administration procedures, for example, the utilization of studbooks and cautious matching proposals, plan to moderate the dangers of inbreeding discouragement.
4. **Restricted Outcome in Renewed introduction:**
   The outcome of renewed introduction endeavors is affected by different elements, including environment reasonableness, predation gambles, and the capacity of delivered people to adjust to wild circumstances. A few animal categories might confront moves in laying out reasonable populaces because of progressing dangers in their normal territories or challenges in adjusting to changed conditions.
5. **Moral Contemplations:**

Hostage rearing projects raise moral contemplations connected with creature government assistance, independence, and the likely effect of bondage on species conduct. Moral conversations include gauging the preservation advantages of hostage rearing against the likely adverse consequences on individual creatures, especially in situations where normal ways of behaving might be compromised.

V. Moral Contemplations in Multiplication and Rearing Projects:

1. **Government assistance of People:**
   The government assistance of individual creatures engaged with proliferation and reproducing programs is an essential moral thought. Hostage conditions ought to focus on the physical and mental prosperity of creatures, guaranteeing that their regular ways of behaving are upheld, and stressors are limited. Sufficient room, fitting social designs, and admittance to appropriate nourishment and veterinary consideration add to the government assistance of people.
2. **Hereditary Contemplations:**
   Adjusting the conservation of hereditary variety with the prosperity of people is a complex moral test. Directors of reproducing programs should settle on informed choices to stay away from impeding impacts of inbreeding while at the same time considering the expected pressure and restrictions forced by imprisonment.

3. **Long haul Supportability:**
   Moral rearing projects think about the drawn out manageability of populaces, both in bondage and, at last, in nature. Administrators should survey the possibility of renewed introduction endeavors and the potential for laying out self-supporting populaces. The objective is to add to the general preservation of species without compromising the prosperity of people.
4. **Straightforwardness and Public Commitment:**

Moral reproducing programs focus on straightforwardness and public commitment. Open correspondence about the objectives, strategies, and difficulties of hostage reproducing cultivates public comprehension and backing. Connecting with the general population in preservation endeavors and giving open doors to schooling add to moral practices in multiplication and rearing projects.

VI. Future Headings and Developments:

1. **Propels in Conceptive Advancements:**
   Progressing research and innovative headways keep on forming the scene of regenerative advancements. Developments like genome altering, high level hormonal treatments, and painless conceptive appraisals offer additional opportunities for overseeing regenerative difficulties in both homegrown and wild species.
2. **Reconciliation of Preservation Advances:**
   The reconciliation of preservation advances, including satellite following, ecological checking, and information examination, improves the viability of rearing projects. These advancements give significant experiences into the way of behaving, wellbeing, and environmental collaborations of delivered people, adding to versatile administration procedures and further developed renewed introduction results.
3. **Cooperative Preservation Endeavors:**
   Coordinated effort between zoos, research establishments, legislative offices, and non-administrative associations is fundamental for the progress of reproducing programs. Joint endeavors work with the trading of information, hereditary material, and mastery, making a cooperative organization zeroed in on protection objectives. Worldwide drives that address natural surroundings protection, against poaching measures, and environmental change add to the more extensive setting of species preservation.
4. **Public Mindfulness and Schooling:**

Public mindfulness and schooling drives assume a critical part in gathering support for protection and moral reproducing rehearses. Zoos, natural life safe-havens, and instructive projects offer open doors for people in general to find out about the significance of biodiversity, the difficulties looked by jeopardized species, and the job of dependable hostage reproducing in preservation endeavors.

6.1 Challenges in captive kangaroo breeding

Hostage kangaroo rearing projects are significant for protection endeavors, examination, and state funded training. Nonetheless, these projects accompany an interesting arrangement of difficulties that should be painstakingly explored to guarantee the prosperity of the kangaroos and the outcome of rearing drives. This investigation digs into the intricacies of hostage kangaroo rearing, addressing difficulties connected with propagation, wellbeing, conduct, and hereditary variety. By getting it and alleviating these difficulties, guardians, veterinarians, and scientists can add to the protection of kangaroo species and the safeguarding of their hereditary variety.

1. **Regenerative Difficulties:**
1. **Occasional Reproducing Examples:**
   Numerous kangaroo species display occasional rearing examples impacted by ecological factors like temperature, precipitation, and food accessibility. In imprisonment, duplicating these regular signs can be testing, prompting hardships in synchronizing rearing cycles. Tending to occasional rearing difficulties requires cautious administration of natural circumstances and regenerative mediations.
2. **Stress-incited Conceptive Concealment:**
   Hostage conditions can prompt pressure in kangaroos, adversely affecting their conceptive wellbeing. Stress-prompted regenerative concealment might bring about unpredictable estrous cycles, diminished ripeness, or the restraint of mating ways of behaving. Establishing low-stress conditions, giving adequate space, and limiting unsettling influences are fundamental techniques to help normal regenerative ways of behaving.
3. **Restricted Reproducing Windows:**
   Kangaroos frequently have restricted windows for fruitful rearing, and botched open doors can affect the generally speaking conceptive outcome of hostage populaces. Cautious checking of estrous cycles, chemical evaluations, and social perceptions are pivotal for recognizing ideal rearing periods. Convenient mediations, like helped regenerative advances, might be important to augment reproducing open doors.
4. **Challenges in Managed impregnation:**

While managed impregnation (computer based intelligence) is a significant device in hostage reproducing, it presents difficulties in kangaroo species. The one of a kind conceptive life systems of female kangaroos, including the bifurcated regenerative parcel, requires particular methods for effective computer based intelligence. Examination and refinement of artificial intelligence conventions are progressing to further develop achievement rates and limit weight on people.

II. Wellbeing Difficulties:

1. **Wholesome Variables:**
   Nourishing uneven characters can fundamentally affect conceptive wellbeing in hostage kangaroos. Lacking eating regimens might prompt lacks in fundamental supplements, influencing richness and the improvement of posterity. Guaranteeing a healthfully adjusted and animal categories proper eating routine is essential for supporting conceptive physiology and advancing generally wellbeing.
2. **Gastrointestinal Issues:**
   Gastrointestinal issues, including parasitic diseases and stomach related messes, can think twice about strength of kangaroos. These issues might affect supplement retention and add to weight reduction, lessening the probability of effective propagation. Routine waste assessments, deworming projects, and dietary changes are fundamental for forestalling and tending to gastrointestinal difficulties.
3. **Metabolic Problems:**
   Metabolic problems, like heftiness and insulin opposition, can emerge in hostage kangaroos because of stationary ways of life and imbalanced weight control plans. These problems might influence conceptive chemicals and lead to sporadic estrous cycles. Carrying out dietary and exercise intercessions to forestall and oversee metabolic problems is basic for conceptive achievement.
4. **Foot and Paw Issues:**

Kangaroos are inclined to foot and paw issues, including pododermatitis, especially when housed on hard surfaces. Foot issues can cause torment and uneasiness, affecting versatility and conceptive ways of behaving. Furnishing substrates with suitable padding, routine foot investigations, and addressing ecological variables add to forestalling and overseeing pododermatitis.

III. Social Difficulties:

1. **Social Elements:**
   Kangaroos are social creatures, and disturbances in friendly elements can affect reproducing ways of behaving. Animosity, contest for assets, or jumbles in matching may prompt pressure and regenerative concealment. Understanding the social designs of kangaroo species is pivotal for establishing agreeable conditions that help normal mating ways of behaving.
2. **Stereotypic Ways of behaving:**
   Hostage kangaroos might foster stereotypic ways of behaving, like pacing or

redundant developments, as a reaction to stress or weariness. Stereotypic ways of behaving can be demonstrative of sub-standard hostage conditions, adversely affecting regenerative achievement. Carrying out ecological advancement, giving adequate room, and advancing normal ways of behaving are fundamental for forestalling stereotypies.

3. **Restricted Space and Fenced in area Plan:**
   Space impediments and unseemly fenced in area configuration can prevent regular regenerative ways of behaving in hostage kangaroos. Lacking space might prompt pressure, decreased active work, and trouble in laying out domains or romance customs. Planning roomy and improving fenced in areas that mirror regular natural surroundings is imperative for supporting kangaroo prosperity and conceptive ways of behaving.
4. **Challenges in Friendly Gathering:**

Making reasonable social groupings is fundamental for kangaroo prosperity and conceptive achievement. Bungles in friendly elements, contrary people, or congestion can prompt pressure and hostility, influencing rearing ways of behaving. Cautious perception, checking social associations, and changing gathering arrangements are pivotal for keeping up with positive social designs.

IV. Hereditary Variety and Populace The executives:

1. **Restricted Hereditary Pool:**
   Hostage kangaroo populaces might confront difficulties connected with a restricted hereditary pool, particularly in little or disconnected rearing gatherings. Inbreeding wretchedness, loss of hereditary variety, and expanded weakness to sicknesses are expected outcomes. Laying out cooperative reproducing programs, hereditary administration techniques, and studbooks are fundamental for keeping up with hereditary variety.
2. **Family Examination:**
   Family examination is a significant instrument for figuring out the hereditary connections inside hostage kangaroo populaces. By following the family and relatedness of people, guardians and geneticists can settle on informed reproducing choices to keep away from inbreeding and advance hereditary variety. Family examination adds to the drawn out wellbeing and maintainability of hostage populaces.
3. **Studbook The board:**
   Studbooks are thorough records of the hereditary history and ancestry of people inside hostage populaces. Studbook the board includes cautious preparation of rearing matches in view of hereditary contemplations, age, and conceptive wellness. Standard updates and cooperation between establishments with shared populaces add to compelling studbook the executives.

4. **Cooperative Rearing Projects:**

To address hereditary difficulties, cooperative rearing projects include associations between foundations, zoos, and untamed life asylums. These projects work with the trading of hereditary material, decrease the dangers of inbreeding, and add to the in general hereditary soundness of hostage kangaroo populaces. Cooperative endeavors upgrade the outcome of rearing drives and backing long haul maintainability.

V. Veterinary Consideration and Conceptive Wellbeing Observing:

1. **Routine Wellbeing Assessments:**
   Routine wellbeing assessments directed by qualified veterinarians are fundamental for checking the general prosperity of hostage kangaroos. These assessments incorporate actual evaluations, regenerative wellbeing checks, and screenings for normal medical problems. Early recognition of regenerative anomalies, wholesome inadequacies, or other wellbeing concerns takes into account ideal mediation.
2. **Hormonal Checking:**
   Hormonal observing assumes a key part in surveying the regenerative soundness of hostage kangaroos. Chemical examines can give experiences into estrous cycles, pregnancy status, and hormonal awkward nature. Observing conceptive chemicals helps with recognizing ideal rearing periods, tending to regenerative difficulties, and supporting helped regenerative innovations.
3. **Ultrasound and Imaging:**
   Ultrasound and imaging procedures are significant devices for regenerative wellbeing checking in kangaroos. These painless techniques permit veterinarians to survey regenerative organs, screen fetal improvement during pregnancy, and distinguish potential issues, for example, pimples or conceptive lot problems. Customary imaging adds to proactive conceptive wellbeing the executives.
4. **Helped Regenerative Advancements:**

Helped regenerative advancements, including managed impregnation and undeveloped organism move, might be utilized to conquer conceptive difficulties in hostage kangaroo rearing. These advances require specific aptitude and cautious coordination between veterinarians, specialists, and overseers. Continuous examination and refinement of strategies add to the progress of helped conceptive projects.

VI. Preservation Training and Public Mindfulness:

1. **Effort and Training Projects:**
   Protection training and public mindfulness programs assume an imperative part in gathering support for hostage kangaroo rearing drives. Zoos, untamed life safe-havens, and instructive establishments can connect with people in general

through outreach programs, directed visits, and instructive materials. These drives assist with bringing issues to light about the significance of hostage rearing in protection and the difficulties looked by kangaroo species.

2. **Local area Inclusion:**
   Including nearby networks in protection endeavors improves public comprehension and backing. Local area commitment projects might incorporate studios, occasions, and cooperative activities that feature the meaning of saving biodiversity, including kangaroo species. Cultivating a feeling of obligation and stewardship among networks adds to the progress of hostage reproducing programs.
3. **Maintainable Practices:**

Preservation schooling stretches out to advancing manageable practices that add to environment conservation and untamed life assurance. Public mindfulness missions can stress the interconnectedness of biological systems and the job of kangaroos in keeping up with natural equilibrium. Empowering economical practices at individual and local area levels upholds more extensive preservation objectives.

6.2Success stories and breakthroughs

Hostage kangaroo rearing projects have seen exceptional examples of overcoming adversity and forward leaps, exhibiting the potential for successful protection endeavors and economical administration of these notorious marsupials. Through the commitment of overseers, veterinarians, and analysts, alongside progressions in helped conceptive advancements and all encompassing administration draws near, a few accomplishments have been acknowledged in hostage kangaroo reproducing.

Conceptive Accomplishment Through Helped Regenerative Advancements (Craftsmanship):

One critical forward leap in hostage kangaroo rearing includes the fruitful utilization of helped conceptive advancements. Planned impregnation (simulated intelligence), undeveloped organism move, and in vitro preparation (IVF) have been utilized to defeat difficulties connected with occasional rearing examples and stress-prompted conceptive concealment. Research establishments and natural life asylums have revealed cases of fruitful pregnancies and births through these methods, adding to the hereditary variety and regenerative outcome of hostage kangaroo populaces.

Hereditary Administration and Studbook Achievement:

The execution of thorough hereditary administration procedures and studbook programs plays had an essential impact in keeping up with sound and hereditarily different hostage kangaroo populaces. Family examination, studbook the board, and cooperative reproducing programs have been fruitful in forestalling inbreeding discouragement and saving the hereditary respectability of kangaroo species. Via cautiously choosing rearing matches in light of hereditary contemplations, guardians have accomplished positive results, guaranteeing the drawn out manageability of hostage populaces.

Effective Renewed introduction Endeavors:

A few hostage rearing projects have made progress in reproducing kangaroos as well as in effectively once again introducing people into their normal living spaces. These renewed introduction endeavors mark a vital achievement in the preservation of kangaroo species, adding to the rebuilding of wild populaces. Examples of overcoming adversity of kangaroos adjusting to their regular habitats, rearing in the wild, and adding to the general environment feature the viability of very much arranged hostage reproducing and renewed introduction drives.

Wellbeing and Government assistance Developments:

Leap forwards in wellbeing and government assistance the executives have fundamentally further developed the general prosperity of hostage kangaroos.

Imaginative ways to deal with sustenance, territory plan, and veterinary consideration have tended to difficulties, for example, foot and paw issues, metabolic problems, and stress-initiated regenerative concealment. Advancement programs that imitate normal ways of behaving, roomy walled in areas, and cooperative endeavors among veterinarians and guardians have added to the physical and mental wellbeing of kangaroo populaces.

Public Mindfulness and Schooling Effect:

Progress in hostage kangaroo rearing stretches out past the bounds of walled in areas. Preservation schooling and public mindfulness programs have taken significant steps in encouraging comprehension and backing for kangaroo protection. Zoos and untamed life asylums have drawn in neighborhood networks through outreach programs, instructive materials, and intuitive displays. These drives not just bring issues to light about the difficulties looked by kangaroos yet in addition move people to take part in preservation endeavors and supportable practices effectively.

Mechanical Advancements in Conceptive Checking:

Mechanical headways have upset regenerative observing in hostage kangaroo rearing projects. Ultrasound imaging, hormonal examines, and painless methods have permitted overseers and veterinarians to intently screen conceptive wellbeing, recognize pregnancies, and survey the suitability of incipient organisms. These advancements give significant bits of knowledge into the regenerative physiology of kangaroos, empowering informed independent direction and proactive administration.

Cooperative Exploration and Worldwide Drives:

Cooperative exploration drives and worldwide organizations have added to the progress of hostage kangaroo rearing projects. Analysts, moderates, and organizations overall team up to share information, hereditary material, and mastery. This cooperative methodology not just improves the hereditary variety of hostage populaces yet additionally fortifies the more extensive protection endeavors pointed toward saving kangaroo species in their regular living spaces.

6.3Ethical considerations in breeding programs

Rearing projects assume a vital part in the protection and safeguarding of imperiled species, both in imprisonment and in nature. Nonetheless, the moral elements of rearing projects are mind boggling, including a fragile harmony between preservation objectives and the government assistance of individual creatures.

This investigation digs into the moral contemplations intrinsic in rearing projects, resolving issues connected with hereditary variety, creature government assistance, renewed introduction endeavors, and the obligation of guardians, veterinarians, and traditionalists.

1. **Hereditary Contemplations:**
1. **Inbreeding and Hereditary Variety:**
   One of the essential moral contemplations in rearing projects is the gamble of inbreeding and the resulting loss of hereditary variety. Inbreeding wretchedness can prompt a scope of medical problems, diminished ripeness, and compromised safe frameworks in posterity. Moral reproducing rehearses focus on hereditary variety, requiring cautious preparation, family investigation, and cooperative endeavors to stay away from the unfortunate results related with inbreeding.
2. **Hereditary Control and Designing:**

The moral ramifications of hereditary control and designing are integral to rearing projects. While innovations like planned impregnation and specific reproducing have been utilized to keep up with hereditary variety, the line becomes obscured while thinking about further developed hereditary mediations. Moral conversations encompass the utilization of advances like quality altering, cloning, and hereditary alteration, raising worries about the expected effect on individual creatures and the more extensive biological system.

II. Creature Government assistance:

1. **Hostage Climate and Personal satisfaction:**
   Guaranteeing the prosperity of creatures in bondage is a foremost moral thought. Rearing projects should focus on establishing conditions that help regular ways of behaving, give adequate room, and advance mental and actual feeling. Moral caretaking includes tending to the mental and actual requirements of people, limiting pressure, and advancing the general personal satisfaction.
2. **Conduct Contemplations:**
   Moral rearing practices perceive and regard the regular ways of behaving of species. In imprisonment, creatures might display stereotypic ways of behaving, showing pressure or disappointment. Moral contemplations include executing improvement programs, planning fenced in areas that emulate regular

territories, and advancing social designs that line up with species-explicit ways of behaving.

3. **Regenerative Wellbeing and Mediation:**

Regenerative wellbeing is a basic part of creature government assistance in reproducing programs. Moral contemplations emerge when mediations are important to beat regenerative difficulties, like helped conceptive innovations or clinical medicines. Finding some kind of harmony between guaranteeing the conceptive progress of an animal types and regarding the independence and prosperity of individual creatures requires cautious moral consideration.

III. Renewed introduction and Preservation:

1. **Achievement Standards for Renewed introduction:**
   Renewed introduction endeavors are a shared objective in reproducing programs meaning to reestablish populaces in their normal environments. Moral contemplations in renewed introduction programs include characterizing achievement rules. It goes past simple populace numbers and incorporates factors, for example, the capacity of delivered people to adjust to the wild, add to biological system elements, and display regular ways of behaving.
2. **Checking and Versatile Administration:**

Moral rearing projects consolidate observing and versatile administration methodologies. Constant appraisal of delivered people and their effect on the environment takes into account acclimations to the program. Moral contemplations stress a guarantee to gaining from encounters, refining draws near, and guaranteeing the drawn out supportability of once again introduced populaces.

IV. Straightforwardness and Public Commitment:

1. **Correspondence of Objectives and Difficulties:**
   Straightforwardness in correspondence is a moral basic in rearing projects. Establishments taking part in reproducing endeavors ought to straightforwardly share their objectives, techniques, and difficulties with the general population. This straightforwardness cultivates trust, instructs the general population about the intricacies of preservation, and empowers support for moral reproducing rehearses.
2. **Public Mindfulness and Instruction:**

Moral reproducing programs perceive the significance of public mindfulness and instruction. Connecting with general society in preservation endeavors, making sense of the meaning of reproducing programs, and giving chances to learning add to informed navigation and a feeling of shared liability regarding biodiversity protection.

V. Cooperative Protection Endeavors:

1. **Worldwide Participation:**
   Moral rearing projects reach out past individual foundations and include worldwide participation. Cooperative endeavors between zoos, research establishments, legislative offices, and non-administrative associations add to shared information, hereditary variety, and protection results. Moral contemplations incorporate a pledge to cooperation for the more extensive objective of species protection.
2. **Impartial Asset Designation:**

Moral contemplations include the impartial designation of assets in reproducing programs. Guaranteeing that assets are dispersed reasonably among species and that the advantages of rearing projects stretch out to the preservation of both notable and less magnetic species is a basic moral aspect.

VI. Long haul Government assistance and Manageability:

1. **Lifetime Care Responsibility:**
   Moral reproducing programs perceive a lifetime care obligation to the creatures in question. This responsibility stretches out past rearing and incorporates arrangements for the prosperity of creatures all through their lives, regardless of whether they are not reasonable for delivery or generation.
2. **Manageable Practices:**

Moral reproducing rehearses accentuate maintainability in all parts of program the board. This incorporates economical reproducing rehearses, preservation drives that think about more extensive biological system wellbeing, and mindful asset use to limit the natural effect of rearing projects.

Chapter 7

Behavioral Observations and Research

Conduct perceptions and exploration are vital parts of grasping the complicated existences of creatures, adding to progressions in ethology, protection, and creature government assistance. This far reaching investigation digs into the meaning of conduct perceptions and examination, featuring their job in disentangling the complexities of creature conduct, advancing preservation endeavors, and upgrading the prosperity of assorted species. From the strategies utilized in conduct exploration to the moral contemplations encompassing the investigation of creature conduct, this assessment plans to give an exhaustive comprehension of the field.

1. **The Significance of Conduct Exploration:**
1. **Experiences into Regular Ways of behaving:**
   Conduct research offers a window into the regular ways of behaving of creatures in their local surroundings. Seeing how creatures connect with their environmental factors, display social ways of behaving, and answer different boosts gives important bits of knowledge into their biological jobs, transformations, and developmental procedures.
2. **Figuring out Species-Explicit Ways of behaving:**
   Various species display interesting ways of behaving formed by their transformative history, biological specialty, and social designs. Social exploration permits researchers to translate species-explicit ways of behaving, going from searching and mating customs to specialized techniques and regional guard. This understanding is major to protection, hostage care, and the executives rehearses.
3. **Commitments to Protection:**
   Conduct research assumes a vital part in protection endeavors, giving fundamental information to the improvement of successful administration procedures. By understanding the way of behaving of imperiled species, analysts can plan preservation designs that address explicit dangers, living space prerequisites, and populace elements. Conduct perceptions help in figuring out procedures

for natural surroundings rebuilding, against poaching endeavors, and renewed introduction programs.

4. **Upgrading Creature Government assistance:**

Conduct research is instrumental in advancing the government assistance of creatures under human consideration. Noticing and deciphering ways of behaving in hostage settings permit overseers and veterinarians to fit conditions to meet the physical and mental necessities of individual creatures. Carrying out advancement exercises, surveying social elements, and addressing stressors add to the general prosperity of creatures in imprisonment.

II. Systems in Conduct Exploration:

1. **Direct Perceptions:**
   Direct perceptions include specialists outwardly archiving and keep creature ways of behaving in their regular territories or controlled conditions. This strategy gives point by point experiences into standards of conduct, social cooperations, and reactions to natural upgrades. Ethologists frequently utilize direct perceptions to construct complete ethograms, classifying an animal categories' collection of ways of behaving.
2. **Remote Detecting Advancements:**
   Progressions in innovation have extended the apparatuses accessible for social exploration. Remote detecting innovations, for example, camera traps, robots, and GPS beacons, empower analysts to notice and gather information on creature conduct without direct human presence. These innovations are especially important for concentrating on subtle or nighttime species.
3. **Bioacoustics and Correspondence Studies:**
   Bioacoustics centers around the investigation of creature sounds and vocalizations. Scientists utilize specific hardware to record and examine the acoustic signs created by creatures, giving bits of knowledge into correspondence, mating calls, and regional presentations. Correspondence studies reach out past vocalizations to incorporate visual signs and non-vocal ways of behaving.
4. **Exploratory Methodologies:**

Exploratory methodologies include controlling factors to notice the subsequent changes in creature conduct. Controlled tests permit analysts to test speculations and draw causal connections between unambiguous elements and social reactions. Moral contemplations are fundamental in trial research, guaranteeing the prosperity of study subjects.

III. Moral Contemplations in Social Exploration:

1. **Creature Government assistance and Sympathetic Treatment:**
   Moral contemplations in social examination focus on the government assistance of study subjects. Scientists should guarantee altruistic treatment, limiting pressure, and keeping away from mischief to creatures during perceptions or examinations. Institutional Creature Care and Use Boards of trustees (IACUCs) assume a vital part in assessing and endorsing research conventions to guarantee moral guidelines are met.
2. **Informed Assent in Non-Human Subjects:**
   In human exploration, informed assent is an essential moral guideline. While creatures can't give express assent, moral contemplations in conduct research underline limiting any expected mischief, regarding normal ways of behaving, and utilizing painless procedures. Analysts should adjust the quest for information with moral treatment.
3. **Long haul Effects and Adjustment:**
   The drawn out effects of conduct research on creature populaces are a basic moral thought. Adjustment, where creatures become acclimated with human presence, can modify regular ways of behaving. Specialists should cautiously oversee adjustment to limit its consequences for concentrated on populaces, particularly with regards to long haul studies.
4. **Protection and Non-Aggravation:**

In protection arranged research, the standard of non-aggravation is fundamental. Analysts should endeavor to limit aggravations to regular natural surroundings, rearing locales, and delicate biological systems. Preservation endeavors ought to focus on the assurance of species and biological systems over the quest for logical information.

IV. Utilizations of Conduct Exploration:

1. **Protection Brain science:**
   Protection brain science applies conduct research standards to grasp human mentalities, convictions, and ways of behaving in regards to preservation endeavors. By looking at the mental variables that impact protection choices, analysts can plan viable effort programs, advance economical ways of behaving, and upgrade public help for preservation drives.
2. **Hostage Creature The board:**
   Social examination is major to the administration of creatures in bondage. Perceptions of hostage conduct illuminate the plan regarding fenced in areas, dietary plans, and enhancement exercises.
   Understanding the social requirements of hostage creatures adds to the making of conditions that advance regular ways of behaving and ideal prosperity.
3. **Human-Natural life Compromise:**
   Social exploration is applied in settling clashes among people and natural life. By

concentrating on the way of behaving of the two people and creatures in struggle inclined regions, scientists can foster methodologies to relieve clashes, diminish pessimistic associations, and advance concurrence. This interdisciplinary methodology tends to the main drivers of contentions and cultivates practical arrangements.

4. **Recovery and Delivery Projects:**

Conduct research advises the improvement regarding recovery and delivery programs for creatures saved from nature. Figuring out normal ways of behaving, social designs, and environmental jobs helps rehabilitators get ready creatures for effective renewed introduction. Social perceptions add to the appraisal of a singular's status for discharge.

V. Challenges in Conduct Exploration:

1. **Onlooker Predisposition and Humanoid attribution:**
   Onlooker predisposition, where scientists inadvertently project their assumptions onto noticed ways of behaving, is a typical test. Humanoid attribution, ascribing human qualities to creatures, can slant understandings. Analysts utilize thorough procedures, between spectator unwavering quality checks, and measurable examinations to relieve these inclinations.
2. **Intricacy of Creature Conduct:**
   Creature conduct is innately perplexing, impacted by a horde of variables. Concentrating on conduct in normal settings requires thinking about factors like ecological circumstances, social elements, and individual variety. The diverse idea of conduct presents difficulties in disconnecting explicit factors and making conclusive determinations.
3. **Restricted Comprehension of Mental Cycles:**
   While recognizable ways of behaving give significant bits of knowledge, the mental cycles hidden conduct stay testing to understand completely. Figuring out creature cognizance, feelings, and abstract encounters requires imaginative techniques and interdisciplinary methodologies, introducing progressing difficulties in conduct research.
4. **Communications in Multi-Species Frameworks:**

Concentrating on ways of behaving in multi-species environments presents extra difficulties. Connections between various species can be perplexing, with flowing impacts on conduct. Research in such frameworks requires a comprehensive methodology that thinks about the environmental setting, interspecies connections, and the powerful idea of biological systems.

VI. Future Bearings in Social Exploration:

1. **Joining of Innovation and Information Examination:**
   The joining of innovation, including man-made consciousness and AI, holds guarantee for progressing social exploration. Mechanized information assortment, refined examination, and example acknowledgment apparatuses empower analysts to deal with huge datasets and uncover nuanced designs in creature conduct. These advancements upgrade effectiveness and extend the extent of examination prospects.
2. **Cross-Disciplinary Cooperation:**
   Future headings in social examination include expanded joint effort across disciplines. Coordinating information from fields like biology, hereditary qualities, neuroscience, and brain science upgrades the profundity of understanding. Cross-disciplinary cooperation cultivates an all encompassing way to deal with concentrating on conduct, taking into account both general and extreme clarifications.
3. **Longitudinal Investigations and Worldwide Data sets:**
   Longitudinal investigations, directed overstretched periods, offer experiences into the elements of creature conduct across various life stages and ecological changes. The making of worldwide data sets works with the sharing of social information, empowering specialists overall to get to data and add to an aggregate comprehension of creature conduct.
4. **Applied Conduct Mediations:**

The utilization of social exploration in protection mediations is a promising road. Conduct mediations can address explicit difficulties, for example, decreasing human-untamed life clashes, advancing feasible ways of behaving, and upgrading the progress of restoration and delivery programs. Applied research adds to prove based preservation rehearses.

7.1 Studying natural behavior in captivity

Concentrating on regular conduct in imprisonment is a basic undertaking that overcomes any barrier between logical figuring out, creature government assistance, and preservation endeavors.

As additional species wind up living in oversaw conditions because of elements like territory misfortune, hostage rearing projects, and exploration drives, the need to grasp and work with normal ways of behaving turns out to be progressively essential. This investigation dives into the difficulties and open doors related with concentrating on regular conduct in hostage settings, accentuating the moral contemplations that guide research rehearses.

1. **Reasoning for Concentrating on Regular Conduct in Bondage:**
1. **Preservation Suggestions:**
   The investigation of regular conduct in imprisonment has direct ramifications

for protection endeavors. As numerous species face dangers in the wild, understanding their regular ways of behaving in hostage settings can illuminate methodologies for environment reclamation, renewed introduction projects, and hostage reproducing drives pointed toward reinforcing wild populaces.

2. **Creature Government assistance Contemplations:**
   Improving the prosperity of creatures in imprisonment requires a far reaching comprehension of their regular ways of behaving. Noticing and concentrating on how creatures express their natural ways of behaving permits overseers and specialists to establish conditions that help physical and mental wellbeing, decrease pressure, and advance species-explicit exercises.
3. **Social Environment Bits of knowledge:**

Concentrating on regular conduct in bondage gives bits of knowledge into the social nature of species. This incorporates grasping social designs, specialized techniques, mating ceremonies, searching ways of behaving, and reactions to different natural boosts. Such information adds to a more all encompassing cognizance of the species' environmental jobs and variations.

II. Challenges in Concentrating on Regular Conduct in Bondage:

1. **Modified Ecological Setting:**
   Hostage conditions, regardless of how very much planned, innately vary from normal living spaces. Creatures in imprisonment might confront imperatives connected with space, substrate, environment, and natural intricacy. These distinctions can impact regular ways of behaving, representing a test for scientists meaning to concentrate on ways of behaving that are setting subordinate.
2. **Hostage Instigated Ways of behaving:**
   Bondage can evoke ways of behaving that are not seen in the wild, usually known as hostage actuated ways of behaving or stereotypies.
   These ways of behaving, like pacing, dull developments, or self-preparing extravagantly, are much of the time indications of stress or dissatisfaction. Recognizing normal ways of behaving and those initiated by bondage is vital for exact translation.
3. **Social Elements and Gathering Design:**
   Hostage settings might present difficulties in duplicating regular social designs and collective vibes. In the wild, creatures frequently display perplexing social ways of behaving impacted by variables like progressive system, participation, and contest. Imprisonment can modify these elements, influencing the outflow of regular ways of behaving inside gatherings.
4. **Human Presence and Onlooker Impacts:**

The simple presence of people can impact creature conduct in bondage. Onlooker impacts, where creatures modify their way of behaving because of the presence of analysts or guardians, should be thought of. Limiting the effect of human presence on noticed ways of behaving is critical for acquiring precise information on regular ways of behaving.

III. Techniques for Concentrating on Regular Conduct in Bondage:

1. **Direct Perceptions:**
   Direct perceptions include deliberately watching and reporting creature conduct continuously. Scientists utilize laid out ethograms, recording an animal categories' collection of ways of behaving, to direct perceptions. Direct perceptions consider the recording of regular ways of behaving, cooperations, and reactions to upgrades in a controlled setting.
2. **Social Studies and Time-Financial plan Investigation:**
   Social studies and time-financial plan investigation include methodicallly recording the event and term of explicit ways of behaving throughout assigned time spans. Time-spending plan investigation gives bits of knowledge into the extent of time creatures apportion to various exercises, revealing insight into their normal social needs.
3. **Ecological Improvement and Naturalistic Walled in areas:**
   Ecological improvement includes altering hostage conditions to empower normal ways of behaving. This can incorporate giving upgrades like articles, substrates, or food astounds that evoke species-explicit ways of behaving. Naturalistic nooks mean to imitate parts of the creatures' normal territory, taking into account more regular articulation of ways of behaving.
4. **Harmless Innovations:**

Progressions in innovation offer harmless ways of concentrating on regular way of behaving. Camera traps, sensors, and wearable gadgets outfitted with accelerometers or GPS trackers take into consideration remote checking of creatures. These innovations give important information while limiting unsettling influence to the subjects.

IV. Moral Contemplations in Concentrating on Regular Conduct in Imprisonment:

1. **Adjusting Exploration Objectives and Creature Government assistance:**
   Moral contemplations in concentrating on normal conduct in imprisonment require a cautious harmony between accomplishing research objectives and guaranteeing the government assistance of the review subjects. Analysts should focus on the prosperity of creatures, limiting pressure, and staying away from any mischief related with the review.

2. **Informed Assent in Creature Exploration:**
   While creatures can't give unequivocal assent, moral contemplations include regarding their independence and limiting any possible mischief or misery. Scientists should comply with moral rules, acquiring essential endorsements from Institutional Creature Care and Use Councils (IACUCs), and guaranteeing that examination conventions focus on creature government assistance.
3. **Limiting Spectator Predisposition and Humanoid attribution:**
   Moral observational examination requires limiting onlooker predisposition and humanoid attribution. Analysts should utilize thorough procedures, between eyewitness unwavering quality checks, and preparing to guarantee that perceptions are level headed and liberated from human understanding inclinations.
4. **Straightforwardness and Detailing Adverse Outcomes:**

Moral exploration rehearses include straightforwardness in detailing both positive and adverse outcomes. Recognizing constraints, difficulties, and occasions where normal ways of behaving may not be completely communicated in that frame of mind to the logical uprightness of the examination and illuminates future examinations.

V. Applications and Ramifications of Concentrating on Regular Conduct in Imprisonment:

1. **Protection Applications:**
   Experiences acquired from concentrating on regular conduct in imprisonment have direct applications in protection. Understanding the social requirements of species in bondage illuminates preservation techniques, including natural surroundings rebuilding, hostage rearing projects, and the improvement of renewed introduction drives.
2. **Hostage Creature The board:**
   Concentrating on regular way of behaving adds to viable hostage creature the board. Perceptions educate the plan regarding nooks, dietary plans, and advancement exercises that help normal ways of behaving and advance the general prosperity of creatures in imprisonment.
3. **Schooling and Public Mindfulness:**
   Perceptions of normal conduct in hostage settings act as significant apparatuses for schooling and public mindfulness. Zoos, untamed life asylums, and instructive organizations can utilize these perceptions to draw in the general population, encouraging comprehension and appreciation for the normal ways of behaving of different species.
4. **Working on Creature Government assistance:**

The down to earth utilizations of concentrating on regular conduct in bondage stretch out to working on creature government assistance. By perceiving and working with normal ways of behaving, guardians and veterinarians can establish conditions that add to the physical and mental prosperity of creatures under human consideration.

VI. Future Headings in Concentrating on Normal Conduct in Imprisonment:

1. **Interdisciplinary Exploration:**
   Future headings include embracing interdisciplinary ways to deal with concentrating on normal conduct in imprisonment. Cooperation between ethologists, environmentalists, veterinarians, and specialists in creature government assistance upgrades the profundity and expansiveness of exploration, consolidating assorted viewpoints and procedures.
2. **Progressions in Harmless Advances:**
   The proceeded with advancement of harmless innovations holds guarantee for propelling examination in concentrating on regular way of behaving. Advancements in camera traps, sensors, and biotelemetry gadgets will furnish analysts with progressively refined devices for observing creatures without direct impedance.
3. **Longitudinal Examinations and Relative Investigations:**
   Longitudinal examinations, directed overstretched periods, offer experiences into changes in normal conduct across various life stages and natural circumstances. Similar examinations, contrasting ways of behaving across related species in various settings, add to a more nuanced comprehension of species-explicit ways of behaving.
4. **Training and Effort Drives:**

Future bearings include fortifying schooling and effort drives. Imparting research discoveries, challenges, and moral contemplations to the public cultivates informed independent direction and backing for preservation endeavors. Interfacing general society with the regular ways of behaving of creatures in imprisonment advances a feeling of stewardship and obligation.

7.2Research contributions to kangaroo conservation

Kangaroos, notable marsupials local to Australia, face different preservation challenges in the cutting edge time. Broad examination endeavors have been fundamental to understanding and tending to these difficulties, contributing altogether to kangaroo protection. This investigation dives into the exploration commitments that have formed kangaroo preservation endeavors, featuring headways, continuous difficulties, and expected future bearings.

1. **Headways in Populace Elements Exploration:**
1. **Understanding Populace Patterns:**

Research on kangaroo populace elements has given pivotal bits of knowledge into the overflow, conveyance, and segment design of various species. Populace concentrates on assist scientists with surveying the wellbeing of kangaroo populaces, distinguish expected dangers, and illuminate protection systems.

2. **Effect of Ecological Elements:**
   Researching the impact of ecological variables on kangaroo populaces has been a key concentration. Research has investigated the effect of environmental change, living space misfortune, and human exercises on kangaroo natural surroundings. Understanding these impacts considers the advancement of designated preservation measures to alleviate adverse consequences.
3. **Hereditary Variety and Protection Hereditary qualities:**

Protection hereditary qualities research plays had an essential impact in surveying the hereditary variety of kangaroo populaces. Understanding hereditary inconstancy is vital for keeping up with sound populaces and forestalling inbreeding. Hereditary examinations add to the improvement of viable rearing and the executives methodologies in imprisonment and nature.

II. Conduct Biology and Protection Systems:

1. **Grasping Regular Ways of behaving:**
   Conduct biology research has given significant experiences into the regular ways of behaving of kangaroos. Perceptions of social designs, mating ways of behaving, and taking care of examples add to the improvement of protection systems that think about the species' environmental jobs and conduct needs.
2. **Human-Natural life Collaborations:**
   Research has investigated the collaborations among kangaroos and human exercises, especially in regions where urbanization infringes upon regular territories. Understanding these communications is urgent for executing measures to limit clashes, decrease vehicle crashes, and advance concurrence among kangaroos and human networks.
3. **Territory The board and Reclamation:**

Concentrates on territory the board and reclamation have informed protection endeavors pointed toward saving and reestablishing kangaroo natural surroundings. Research discoveries guide the plan of preservation programs that emphasis on keeping up with practical biological systems, safeguarding biodiversity, and guaranteeing the accessibility of reasonable natural surroundings for kangaroo populaces.

III. Wellbeing and Illness The executives:

1. **Illness Observation and Checking:**
   Research on the strength of kangaroo populaces includes infection reconnaissance and observing. Understanding the commonness of infections and their

effect on kangaroo populaces is basic for executing proactive measures to forestall and oversee likely flare-ups, guaranteeing the drawn out strength of the species.

2. **Veterinary Intercessions:**
   Veterinary examination has added to the advancement of mediations to address medical problems in both hostage and wild kangaroo populaces. Viable sickness the executives, therapy conventions, and restoration procedures are educated by research discoveries, adding to worked on generally speaking wellbeing and government assistance.
3. **One Wellbeing Approaches:**

Research has progressively embraced "One Wellbeing" approaches, perceiving the interconnectedness of human, creature, and natural wellbeing.

This all encompassing point of view thinks about the more extensive environment and the ramifications of kangaroo wellbeing on biological system wellbeing. One Wellbeing research illuminates far reaching protection methodologies that address various features of the biological system.

IV. Preservation Instruction and Public Mindfulness:

1. **Correspondence Procedures:**
   Research has investigated successful correspondence procedures to bring issues to light about kangaroo protection. Grasping public insights, mentalities, and information holes considers the advancement of designated instruction programs. Research-informed correspondence procedures mean to connect with general society in protection endeavors and encourage a feeling of obligation.
2. **Local area Commitment Projects:**
   Local area commitment research adds to the improvement of projects that include neighborhood networks in kangaroo preservation. By cultivating a feeling of stewardship and cooperation, local area commitment drives upgrade the viability of preservation gauges and empower support for kangaroo security.
3. **Moral Contemplations in Protection Informing:**

Research has dug into moral contemplations in preservation informing, guaranteeing that instructive endeavors are exact, aware, and socially touchy. Moral correspondence methodologies add to building positive connections between preservation associations, analysts, and the networks impacted by kangaroo protection endeavors.

V. Challenges in Kangaroo Preservation Exploration:

1. **Restricted Subsidizing and Assets:**
   In spite of the meaning of kangaroo protection research, restricted financing and assets present continuous difficulties. Getting sufficient financing is essential for

supporting long haul studies, hands on work, and the execution of protection drives.

2. **Intricacies in Human-Natural life Compromise:**
   Human-natural life clashes, especially in regions where kangaroos and human exercises cross-over, present complex difficulties. Research is expected to foster compelling compromise methodologies that balance the necessities of both kangaroo populaces and human networks.
3. **Adjusting to Environmental Change:**

Environmental change represents a huge danger to kangaroo living spaces and biological systems. Research is fundamental to comprehend how kangaroo populaces are adjusting to changing climatic circumstances and to foster preservation techniques that alleviate the effect of environmental change on their natural surroundings.

VI. Future Headings in Kangaroo Protection Exploration:

1. **Innovation and Information Examination:**
   Progressions in innovation, including satellite symbolism, remote detecting, and information examination, offer new roads for kangaroo preservation research. These instruments empower scientists to assemble and dissect huge data-sets, giving a more thorough comprehension of kangaroo populaces and their territories.
2. **Local area Based Protection Drives:**
   Future examination can zero in on local area based preservation drives that effectively include neighborhood networks in kangaroo protection. Understanding the socio-social elements and teaming up with networks can prompt more viable and manageable protection procedures.
3. **Worldwide Coordinated effort and Information Sharing:**
   Worldwide coordinated effort and information sharing are urgent for tending to protection challenges that stretch out past public lines. Research organizations, cooperative ventures, and information sharing drives work with the trading of data and best works on, adding to a worldwide comprehension of kangaroo preservation.
4. **Coordinated Preservation Arranging:**

Coordinated protection arranging, integrating research from assorted disciplines, can upgrade the viability of preservation endeavors. Interdisciplinary methodologies that think about biological, social, and financial variables add to complete protection systems that advance the drawn out manageability of kangaroo populaces.

7.3Ethical considerations in behavioral studies

Social examinations, whether directed in the wild or bondage, include the unpredictable investigation of creature conduct to develop how we might interpret species-explicit attributes, social elements, and environmental jobs. Be that as it may, the quest for information in social examinations should be morally directed, recognizing the government assistance and privileges of the review subjects. Adjusting the obtaining of significant experiences with moral contemplations is principal in guaranteeing the dependable direct of examination.

1. **Regard for Creature Government assistance:**
   Moral contemplations in social examinations focus on the government assistance of study subjects. Analysts should limit pressure, distress, and any mischief that could result from the review. The execution of painless strategies, adherence to laid out moral rules, and the oversight of Institutional Creature Care and Use Councils (IACUCs) are fundamental in maintaining the prosperity of creatures engaged with conduct research.
2. **Informed Assent in Non-Human Subjects:**
   While human investigations require informed assent, creatures can't give unequivocal assent. Moral contemplations in social examinations include regarding the independence of the review subjects to the degree conceivable. Analysts should cautiously configuration studies to limit impedance with normal ways of behaving and abstain from inciting pressure, guaranteeing that the noticed ways of behaving are legitimate portrayals.
3. **Limiting Spectator Inclination and Humanoid attribution:**
   Moral exploration rehearses request objectivity in perceptions. Spectator inclination, where specialists accidentally project their assumptions onto noticed ways of behaving, and humanoid attribution, ascribing human qualities to creatures, should be limited. Thorough approaches, between eyewitness unwavering quality checks, and preparing programs assist with guaranteeing that translations are grounded in exact proof as opposed to human points of view.
4. **Straightforwardness in Revealing:**
   Moral social examinations require straightforwardness in announcing discoveries, including both positive and adverse outcomes. Straightforward revealing adds to the logical trustworthiness of the exploration and takes into consideration an extensive comprehension of the review's results. Recognizing constraints, difficulties, and occasions where regular ways of behaving may not be completely communicated is pivotal in advancing moral exploration rehearses.
5. **Long haul Effects and Adjustment:**
   Thought of the drawn out effects of conduct concentrates on creature populaces is a moral objective. Adjustment, where creatures become acquainted with human presence, can modify normal ways of behaving. Analysts should oversee adjustment cautiously to limit its belongings, particularly with regards to long haul studies, and focus on the preservation of normal ways of behaving inside populaces.

6. **Preservation and Non-Aggravation:**

Social investigations frequently add to protection endeavors, however moral contemplations include limiting unsettling influences to regular natural surroundings, reproducing destinations, and delicate biological systems. Preservation situated exploration ought to focus on the security of species and biological systems over the quest for logical information, encouraging a pledge to non-unsettling influence.

Chapter 8

Educational Outreach and Public Awareness

Instructive effort and public mindfulness assume crucial parts in cultivating grasping, appreciation, and dynamic support in preservation endeavors. As ecological difficulties escalate, the requirement for educated and connected with networks turns out to be progressively basic. This exhaustive investigation dives into the complex domain of instructive effort and public mindfulness, looking at their significance, procedures for viable execution, contextual analyses, challenges, and the developing scene of natural schooling.

1. **The Meaning of Instructive Effort and Public Mindfulness:**
1. **Enabling Informed Independent direction:**
   Instructive effort outfits people with the information and apparatuses to settle on informed choices in regards to ecological issues. By giving exact data about biological systems, biodiversity, and preservation rehearses, networks are enabled to add to maintainable arrangements and promoter for dependable natural stewardship.
2. **Encouraging a Preservation Ethic:**
   Public mindfulness drives add to the improvement of a protection ethic — a common feeling of obligation and regard for the climate. Developing a profound association with nature and a comprehension of the reliance of biological systems urges people to embrace ways of behaving that limit ecological effect and add to the protection of regular assets.
3. **Building a Maintainable Future:**

Instructive effort and public mindfulness are basic parts of building a maintainable future. By imparting natural education, cultivating an appreciation for biodiversity, and advancing economical practices, these drives lay the foundation for networks to effectively take part in making and keeping a decent and strong climate.

II. Systems for Successful Instructive Effort:

1. **School-Based Projects:**
   Carrying out natural training in schools gives an organized stage to arriving at youthful personalities. Curricular incorporation, open air opportunities for growth, and extracurricular exercises add to building a groundwork of ecological mindfulness among understudies. Instructive foundations assume a critical part in molding the qualities and mentalities of people in the future.
2. **Local area Studios and Courses:**
   Facilitating studios and courses inside networks fills in as a powerful method for contacting a different crowd. Points can go from reasonable living practices and untamed life protection to environmental change moderation. Drawing in specialists, teachers, and local area pioneers upgrades the believability and effect of such occasions.
3. **Nature-Based Encounters:**
   Vivid encounters in nature, like directed climbs, untamed life safaris, and eco-visits, give active learning open doors. Interfacing individuals straightforwardly with common habitats cultivates a special interaction to the regular world, empowering a feeling of obligation and stewardship.
4. **Online Stages and Advanced Media:**

Utilizing on the web stages and computerized media grows the span of instructive effort. Sites, web-based entertainment crusades, web recordings, and online classes empower associations to spread data generally and draw in with a worldwide crowd. Computerized media takes into consideration intuitive and dynamic substance conveyance, taking special care of different learning inclinations.

III. Contextual analyses in Fruitful Instructive Effort:

1. **Jane Goodall Organization's Foundations and Shoots Program:**
   The Roots and Shoots program, started by eminent primatologist Jane Goodall, represents effective instructive effort. Working in more than 140 nations, the program connects with youth in natural and philanthropic ventures, cultivating a feeling of obligation and activism. Through a blend of school-based exercises and local area projects, Roots and Shoots engages youthful pioneers to drive positive change.
2. **Earthwatch Organization's Resident Science Projects:**
   The Earthwatch Organization stresses resident science as a useful asset for instructive effort. By including volunteers in logical examination projects, Earthwatch contributes significant information to natural investigations as well as furnishes members with firsthand openness to protection challenges. Resident science programs improve public comprehension and commitment to logical undertakings.
3. **Public Geographic's Schooling Drives:**

Public Geographic's schooling drives incorporate a large number of assets, including illustration plans, intuitive guides, and virtual encounters. These drives influence the association's tremendous organization of researchers, picture takers, and voyagers to bring genuine investigation into homerooms. By making investigation open and connecting with, Public Geographic moves a feeling of miracle and interest on the planet.

IV. Public Mindfulness Missions:

1. **World Untamed life Asset's "Save Tigers Currently" Mission:**
   The World Natural life Asset's "Save Tigers Presently" crusade is an outstanding illustration of a public mindfulness drive. Consolidating VIP supports, virtual entertainment commitment, and on-the-ground preservation endeavors, the mission effectively brought issues to light about the predicament of tigers and collected help for tiger protection. Public mindfulness missions of this nature influence media and mainstream society to contact expansive crowds and drive aggregate activity.
2. **The Sierra Club's "Past Coal" Mission:**
   The Sierra Club's "Past Coal" crusade epitomizes how designated public mindfulness missions can impact strategy and general assessment. Zeroed in on changing away from coal-based energy, the mission uses grassroots getting sorted out, narrating, and support to illuminate networks about the ecological and wellbeing effects of coal and backer for cleaner energy choices.
3. **The Plastic Contamination Alliance's Web-based Entertainment Missions:**

Web-based entertainment has turned into a useful asset for public mindfulness crusades, as exhibited by the Plastic Contamination Alliance. Through significant visuals, infographics, and intuitive substance, the association brings issues to light about the adverse impacts of plastic contamination. Virtual entertainment crusades welcome people to take part in lessening plastic utilization and promoter for feasible other options.

V. Challenges in Instructive Effort and Public Mindfulness:

1. **Defeating Data Over-burden:**
   In a world immersed with data, slicing through the clamor to pass on significant natural messages represents a test. Instructive effort programs should utilize powerful correspondence procedures to catch and hold the crowd's consideration in the midst of contending needs.
2. **Social and Phonetic Variety:**
   Different social and phonetic foundations inside networks require nuanced ways to deal with instructive effort. Messages and materials ought to be socially

delicate, made an interpretation of fittingly, and customized to reverberate with the qualities and points of view of various segment gatherings.

3. **Estimating Effect and Social Change:**

Assessing the effect of instructive effort programs and surveying conduct change among members can challenge. Estimating long haul influences, like changes in perspectives and supported harmless to the ecosystem rehearses, requires vigorous assessment strategies and follow-up examinations.

VI. Future Headings in Instructive Effort and Public Mindfulness:

1. **Computer generated Reality (VR) and Increased Reality (AR):**
   The mix of computer generated reality (VR) and expanded reality (AR) advancements holds guarantee for vivid instructive encounters. VR and AR can move people to virtual biological systems, furnishing sensible and significant experiences with nature and untamed life, in this way upgrading ecological schooling.
2. **Gamification for Learning:**
   Gamification, integrating game components into instructive substance, presents a drawing in approach for ecological training. Instructive games and applications can make learning fun, intelligent, and open to a wide crowd, especially more youthful ages who are familiar with computerized interfaces.
3. **Local area Based Participatory Exploration (CBPR):**
   Local area based participatory exploration includes effectively captivating networks in the examination cycle. This approach guarantees that instructive effort drives are socially pertinent, address nearby worries, and engage networks to add to preservation endeavors effectively.
4. **Interdisciplinary Methodologies:**

Future instructive effort drives can profit from interdisciplinary joint efforts, coordinating bits of knowledge from fields like brain science, humanism, and correspondences. Grasping the mental variables that impact conduct, cultural perspectives, and viable correspondence procedures upgrades the adequacy of effort endeavors.

8.1Role of captive kangaroos in education

Hostage kangaroos, living in zoos, natural life safe-havens, and instructive organizations, assume a multi-layered part in overcoming any barrier between logical information and public comprehension. Past being enthralling shows, these marsupials act as diplomats for their wild partners, offering extraordinary open doors for instruction, exploration, and cultivating preservation mindfulness. This investigation digs into the different manners by which hostage kangaroos add to schooling, stressing their part in advancing natural education, logical request, and a more profound appreciation for biodiversity.

1. **Ecological Proficiency and Biodiversity Schooling:**
1. **Species-Explicit Schooling:**
   Hostage kangaroos give a substantial association with Australia's remarkable environments and biodiversity. Instructive projects based on these marsupials offer experiences into their way of behaving, living space inclinations, and the job they play in their local surroundings. Such projects add to a more extensive comprehension of the interconnectedness of biological systems and the significance of protecting biodiversity.
2. **Protection Informing:**
   Hostage kangaroos act as diplomats for their wild partners, empowering teachers to pass on protection messages really. Finding out about the preservation status, dangers, and progressing endeavors to safeguard kangaroos in the wild imparts a feeling of obligation and energizes proactive protection activities among the crowd.
3. **Advancing Maintainable Practices:**

Instruction worked with by hostage kangaroos stretches out past the actual species. Zoos and natural life asylums frequently incorporate data about supportable practices, territory protection, and the more extensive effect of human exercises on natural life. Guests gain bits of knowledge into the significance of settling on ecologically cognizant decisions to help the prosperity of kangaroos as well as the biological systems they possess.

II. Logical Schooling and Request:

1. **Observational Learning Open doors:**
   Zoos and untamed life asylums give interesting observational learning open doors to understudies and guests. Noticing hostage kangaroos takes into account the firsthand investigation of their way of behaving, social elements, and variations. Such encounters animate interest and energize logical request among students, all things considered.
2. **Intelligent Opportunities for growth:**
   Intelligent projects, like directed visits, creature experiences, and instructive showings, take into consideration involved opportunities for growth with hostage kangaroos. These projects take care of various learning styles and make vital encounters that leave an enduring effect, encouraging a deep rooted interest in science, zoology, and ecological sciences.
3. **Coordinated effort with Instructive Organizations:**

Numerous zoological organizations team up with schools and colleges to improve formal instruction. Hostage kangaroos become subjects of exploration projects, contributing significant information to logical investigations. These joint efforts furnish understudies with functional exploration experience, enhancing how they might interpret creature conduct, physiology, and preservation science.

III. Public Commitment and Effort:

1. **Guest Commitment Projects:**
   Zoos and untamed life safe-havens foster guest commitment programs that influence hostage kangaroos as instructive resources. Interpretive signage, directed visits, and educational meetings assist guests with interfacing with the kangaroos on a more profound level, encouraging a feeling of miracle and appreciation for these famous marsupials.
2. **Unique Occasions and Mindfulness Missions:**
   Extraordinary occasions and mindfulness crusades revolved around hostage kangaroos cause to notice explicit protection issues. These drives might harmonize with public or worldwide untamed life mindfulness days, permitting foundations to teach people in general about kangaroo preservation, natural surroundings assurance, and the worldwide meaning of biodiversity.
3. **Instructive Studios and Showings:**

Instructive studios and showings give a stage to top to bottom finding out about kangaroos and their environments. These projects might incorporate introductions by untamed life specialists, intelligent exhibitions, and open doors for crowd investment. By joining amusement with training, these drives catch the interest of different crowds.

IV. Local area Effort and Inclusivity:

1. **Nearby People group Inclusion:**
   Zoos and natural life asylums frequently stretch out their instructive effort to nearby networks. Designated programs plan to include occupants in protection drives, featuring the job of hostage kangaroos in cultivating a deep satisfaction and obligation regarding neighborhood biodiversity.
2. **Inclusivity and Availability:**
   Instructive drives based on hostage kangaroos make progress toward inclusivity and openness. Outreach projects might be intended to oblige different crowds, incorporating people with inabilities, to guarantee that everybody has the amazing chance to draw in with and gain from these astounding marsupials.
3. **Organizations with Native People group:**

Joint efforts with Native people group add to a more all encompassing comprehension of kangaroos inside their social setting. Instructive projects that integrate Native points of view on kangaroos and their importance improve the social wealth of the instructive experience, advancing multifaceted comprehension and regard.

V. Challenges and Moral Contemplations:

1. **Adjusting Schooling and Creature Government assistance:**
   One of the essential difficulties in using hostage kangaroos for schooling is finding some kind of harmony between instructive objectives and creature government assistance. Establishments should guarantee that instructive projects are planned with the prosperity of the kangaroos as a main concern, limiting pressure and giving proper conditions.
2. **Guest Confusion:**
   Guaranteeing that guests get precise and science-based data is vital. Error of kangaroo conduct or nature might prompt misguided judgments. Instructive projects should be cautiously arranged to convey logical information and disperse normal fantasies or mistaken assumptions.
3. **Beating Sentimentality:**

The appealling idea of kangaroos may at times prompt sentimentality, eclipsing instructive messages. Organizations should be cautious in introducing real data and keeping away from the utilization of misrepresented accounts that might think twice about instructive uprightness of projects.

VI. Future Headings in Kangaroo Training:

1. **Progressions in Advanced Learning:**
   Advanced stages and virtual encounters offer open doors for extending kangaroo training past actual displays. Computer generated reality (VR) encounters, online instructive modules, and intuitive applications can improve availability and give vivid learning valuable open doors to worldwide crowds.
2. **Reconciliation of Native Information:**
   Future instructive drives can additionally coordinate Native information and viewpoints into kangaroo training. Joint efforts with Native people group can improve instructive substance, giving a more complete comprehension of the social, biological, and verifiable meaning of kangaroos.
3. **Worldwide Joint effort for Protection Instruction:**

Cooperative endeavors among zoos, untamed life safe-havens, and instructive establishments on a worldwide scale can improve the effect of kangaroo schooling. Shared assets, cooperative examination tasks, and joint mindfulness crusades add to a brought together methodology in cultivating worldwide preservation mindfulness.

8.2Engaging the public in conservation efforts

Protection endeavors are at the front of tending to natural difficulties and safeguarding biodiversity. A urgent part of accomplishing effective preservation results is connecting with people in general as dynamic members in these drives. Public commitment brings issues to light as well as cultivates a feeling of obligation and possession, making an economical association among networks and preservation associations. This investigation digs into the meaning of connecting with general society in preservation endeavors, the different techniques utilized, fruitful contextual investigations, challenges confronted, and future headings for building a vigorous and cooperative protection development.

1. **Meaning of Public Commitment to Preservation:**
1. **Catalyzing Social Change:**
   Public commitment fills in as an impetus for social change. At the point when people comprehend the effect of their activities on the climate, they are bound to embrace economical practices, diminish their biological impression, and backing preservation drives. This social shift is major to the drawn out progress of preservation endeavors.
2. **Making Preservation Supporters:**
   Connected with people frequently become advocates for protection causes. At the point when the general population is educated about ecological issues, they are better prepared to convey the significance of preservation to their friends, families, and networks. This gradually expanding influence intensifies the range and effect of preservation messages.
3. **Adding to Resident Science:**

Public commitment stretches out to resident science drives where people effectively take part in information assortment, checking, and research. These commitments give significant data to preservation projects, improve logical comprehension, and enable residents to assume an immediate part in protecting biodiversity.

II. Procedures for Compelling Public Commitment:

1. **Schooling and Mindfulness Missions:**
   Far reaching schooling and mindfulness crusades are fundamental to public commitment. These missions use different channels, including web-based entertainment, conventional media, and instructive establishments, to disperse data about natural issues, preservation rehearses, and the significance of biodiversity.
2. **Intelligent Projects and Occasions:**
   Facilitating intelligent projects and occasions sets out open doors for direct commitment. Studios, nature strolls, untamed life celebrations, and intelligent shows permit general society to interface with protection subjects on an individual level. Involved encounters cultivate a more profound comprehension and appreciation for the regular world.
3. **Joint efforts with Local area Pioneers:**
   Working together with local area pioneers, powerhouses, and neighborhood

specialists can improve the effect of protection messages. Pioneers who champion natural causes can impact general assessment, advocate for strategy changes, and assemble networks to take part in preservation drives effectively.

4. **Advanced Stages and Virtual Entertainment:**

Utilizing advanced stages and virtual entertainment is instrumental in contacting different crowds worldwide. Drawing in happy, online missions, and intuitive stages work with constant correspondence, empowering protection associations to interface with people of any age and foundations.

III. Contextual analyses in Fruitful Public Commitment:

1. **The Jane Goodall Establishment's Foundations and Shoots Program:**
   The Roots and Shoots program, established by primatologist Jane Goodall, embodies fruitful public commitment. Working in north of 140 nations, the program draws in youth in ecological and philanthropic tasks. Through school-based exercises, local area undertakings, and youth-drove drives, Roots and Shoots enables youngsters to add to protection endeavors effectively.
2. **The Incomparable Boundary Reef Establishment's Resident Science Undertakings:**
   Resident science drives drove by the Incomparable Boundary Reef Establishment have effectively connected with the general population in the protection of coral reefs. Through programs like the "Eye on the Reef" project, people, including travelers and local people, contribute information on reef wellbeing, coral fading, and marine life sightings. This participatory methodology upgrades public comprehension and association in reef protection.
3. **Public Geographic's Planet or Plastic Mission:**

Public Geographic's "Planet or Plastic" crusade is a convincing instance of utilizing media impact for public commitment. Through significant visuals, narratives, and an internet based stage, the mission brings issues to light about the worldwide plastic contamination emergency. It urges people to lessen single-utilize plastic utilization and supporters for foundational changes to resolve the issue.

IV. Challenges in Open Commitment for Preservation:

1. **Beating Detachment and Carelessness:**
   Disregard and lack of concern present huge difficulties in connecting with people in general. A few people might feel separated from natural issues or accept that their activities don't have a significant effect. Beating this latency requires designated informing and showing the unmistakable results of aggregate endeavors.

2. **Exploring Social and Etymological Variety:**
   Social and semantic variety can be a hindrance to successful correspondence. Preservation messages should be socially delicate, made an interpretation of suitably, and customized to reverberate with the qualities and viewpoints of various segment gatherings. Modifying commitment techniques for different networks is critical for inclusivity.
3. **Adjusting Neighborhood and Worldwide Needs:**

Finding some kind of harmony among neighborhood and worldwide preservation needs is intricate. Public commitment systems need to resolve gives that reverberate with nearby networks while accentuating the interconnectedness of worldwide environments. Adjusting worldwide protection objectives to nearby requirements guarantees pertinence and adequacy.

V. Future Headings in Open Commitment for Protection:

1. **Tackling Innovation for Virtual Encounters:**
   Progressions in innovation, including computer generated reality (VR) and expanded reality (AR), offer new roads for public commitment. Virtual encounters can move people to environments, permitting them to investigate and value biodiversity without actual presence. Intelligent computerized stages upgrade openness and give vivid opportunities for growth.
2. **Advancing Economical Ways of life:**
   Future public commitment drives can zero in on advancing maintainable ways of life. Stressing the positive effect of individual decisions, for example, practical utilization, sustainable power use, and preservation disapproved of ways of behaving, urges long haul responsibilities to natural stewardship.
3. **Integrating Native Information and Points of view:**
   Incorporating Native information and points of view into public commitment drives upgrades social wealth and credibility. Coordinated efforts with Native people group add to a more all encompassing comprehension of protection, integrating customary biological information and encouraging admiration for different social ways to deal with manageability.
4. **Local area Based Protection Projects:**

Local area based protection programs engage nearby networks to play a functioning job in preservation. Drives that include occupants in direction, maintainable asset the board, and preservation projects encourage a feeling of pride and obligation to safeguarding neighborhood biodiversity.

8.3Balancing entertainment and education

The crossing point of diversion and schooling has become progressively significant in making connecting with and effective opportunities for growth. Adjusting these two components is a sensitive yet fundamental undertaking, especially in the domains of media, innovation, and instructive settings. This investigation dives into the meaning of adjusting amusement and schooling, the difficulties and advantages intrinsic in this methodology, procedures for accomplishing balance, and instances of effective executions across different stages.

1. **The Meaning of Adjusting Amusement and Instruction:**
1. **Commitment and Maintenance:**
   The combination of amusement into instructive substance upgrades commitment and maintenance. Integrating components, for example, narrating, gamification, and intuitive encounters catches students' consideration, making the instructive material more essential and agreeable. Expanded commitment adds to better maintenance of data.
2. **Cultivating Inborn Inspiration:**
   Amusement has the ability to cultivate characteristic inspiration in students. At the point when instructive substance is introduced in a connecting with and pleasant way, people are bound to be spurred by interest and a certified interest in learning. This inborn inspiration advances an uplifting outlook towards instruction.
3. **Interesting to Assorted Learning Styles:**

People have assorted learning styles, and adjusting amusement and training obliges these distinctions. Visual students might profit from media introductions, while sensation students might flourish in intelligent and active encounters. An even methodology guarantees that instructive substance requests to an expansive scope of students.

II. Challenges in Adjusting Amusement and Training:

1. **Keeping up with Scholastic Meticulousness:**
   One of the difficulties in adjusting amusement and training is keeping up with scholarly thoroughness. While diversion improves commitment, there is a gamble of weakening the instructive substance to the place where it compromises the profundity and intricacy expected for an exhaustive opportunity for growth. Finding some kind of harmony is fundamental to guarantee the two components coincide amicably.
2. **Tending to Limited ability to focus:**
   In a time of data over-burden and limited ability to focus, catching and keeping up with students' center is testing. Adjusting diversion helps address this issue, however there is a scarce difference between connecting with content and shallow commitment. Teachers and content makers should track down ways of conveying significant data while taking special care of abbreviated capacities to focus.

3. **Guaranteeing Precise Data:**

The quest for amusement shouldn't think twice about exactness and validity of instructive substance. Deception can multiply while possibly not cautiously arranged, prompting mistaken assumptions and confusions. Guaranteeing that engaging components don't misshape the instructive message is significant for the honesty of the opportunity for growth.

III. Advantages of Adjusting Amusement and Schooling:

1. **Improved Learning Results:**
   The incorporation of diversion upgrades learning results by making instructive substance more available and agreeable. At the point when students are effectively connected with and spurred, they are bound to hold data and apply it in different settings. Diversion goes about as an impetus for successful growth opportunities.
2. **Long lasting Learning:**
   Adjusting diversion and instruction adds to the improvement of an inspirational perspective towards picking up, encouraging a culture of deep rooted learning. At the point when people find delight and fulfillment in the growing experience, they are bound to search out instructive open doors past proper settings and all through their lives.
3. **Innovative Critical thinking:**

The mix of amusement and instruction sustains innovative reasoning and critical thinking abilities. Intelligent and connecting with instructive encounters urge students to think basically, investigate elective arrangements, and apply information in creative ways. These abilities are fundamental for adjusting to a consistently impacting world.

IV. Procedures for Adjusting Amusement and Schooling:

1. **Integrating Gamification:**
   Gamification includes incorporating game components, like difficulties, prizes, and rivalry, into instructive substance. This technique use the innately engaging parts of games to make learning more pleasant and locking in. Instructive games can cover a great many subjects and take care of different age gatherings.
2. **Intuitive Sight and sound Introductions:**
   Mixed media introductions, including recordings, livelinesss, and recreations, can change instructive substance into outwardly captivating and intelligent encounters. Stages like computer generated simulation (VR) and increased reality (AR) give vivid conditions that improve understanding and maintenance of perplexing ideas.

3. **Narrating and Story driven Learning:**
   Narrating is an amazing asset for meshing instructive substance into a story structure. Making convincing stories around instructive ideas catches students' consideration, summons close to home reactions, and makes the material more engaging. Story driven growth opportunities are especially compelling ever, science, and writing.
4. **True Applications and Encounters:**
   Interfacing instructive substance to certifiable applications and encounters adds a functional aspect to learning. Field trips, involved exercises, and genuine contextual investigations make conceptual ideas unmistakable and important. This approach stresses the reasonable utility of information, expanding its importance for students.
5. **Versatile Learning Stages:**

Versatile learning stages use innovation to customize instructive encounters in view of individual student profiles. These stages break down student progress, adjust content conveyance, and give input customized to every understudy's necessities. Personalization upgrades commitment and permits students to advance at their own speed.

V. Instances of Effective Adjusting of Amusement and Instruction:

1. **Khan Foundation:**
   Khan Foundation is a web-based instructive stage that successfully balances diversion and schooling. Through intuitive recordings, practice works out, and gamified components, Khan Institute makes picking up connecting with and available. The stage covers many subjects and takes special care of students, everything being equal.
2. **TED-Ed:**
   TED-Ed, an augmentation of the TED Talks stage, centers around making short, energized examples that balance diversion and training. The outwardly captivating movements supplement skillfully organized content, making complex subjects more edible for a worldwide crowd.
3. **Minecraft: Instruction Release:**

Minecraft: Instruction Release is an illustration of gamification applied to training. This well known game has an instructive release explicitly intended for study hall use. It permits understudies to team up, issue tackle, and apply inventiveness inside a virtual climate, mixing diversion with learning.

VI. Challenges in Carrying out Adjusting Systems:

1. **Admittance to Innovation:**
   Carrying out systems that balance diversion and training, especially those

including innovation, might be frustrated by inconsistent admittance to computerized assets. Financial variations can bring about certain students having restricted admittance to innovation, affecting their capacity to profit from specific instructive methodologies.

2. **Instructor Preparing and Proficient Turn of events:**
   Teachers might confront difficulties in integrating amusement into their showing techniques, particularly in the event that they need preparing in these creative methodologies. Proficient improvement open doors that furnish instructors with the abilities to adjust diversion and schooling are fundamental for effective execution.
3. **Appraisal and Normalization:**

Conventional evaluation techniques and government sanctioned testing may not really catch the advantages of diversion imbued instructive methodologies. Accommodating the requirement for normalized evaluation with the dynamic and customized nature of engaging growth opportunities represents a test for instructors and instructive establishments.

VII. Future Headings in Adjusting Diversion and Schooling:

1. **Computerized reasoning (simulated intelligence) in Schooling:**
   The mix of computerized reasoning (man-made intelligence) holds guarantee for customized and engaging instructive encounters. Artificial intelligence calculations can adjust content progressively, giving fitted learning pathways to individual understudies. Chatbots and virtual mentors controlled by simulated intelligence can offer quick input and direction.
2. **Vivid Computer generated Simulation (VR) and Increased Reality (AR):**
   Progressions in vivid advances, like computer generated simulation (VR) and expanded reality (AR), will keep on reclassifying instructive encounters. Virtual field trips, intelligent reproductions, and three-layered learning conditions improve commitment and give novel approaches to introducing instructive substance.
3. **Cross-disciplinary Joint efforts:**

Joint efforts between instructors, content makers, and media outlet experts can prompt imaginative cross-disciplinary methodologies. Consolidating the aptitude of teachers with the imagination of narrators, movie producers, and game designers can result in synergistic instructive and amusement encounters.

Chapter 9

Conservation Efforts and Future Directions

Preservation endeavors are basic despite heightening ecological difficulties and biodiversity misfortune. Throughout the long term, devoted people, associations, and state run administrations have worked resolutely to safeguard biological systems, save jeopardized species, and advance manageable practices. This far reaching investigation dives into the present status of worldwide preservation endeavors, featuring effective drives, tending to difficulties confronted, and illustrating future headings to guarantee a versatile and biodiverse planet.

1. **The Present status of Worldwide Preservation Endeavors:**

1. **Safeguarded Regions and Stores:**
   One foundation of preservation endeavors includes laying out and keeping up with safeguarded regions and stores. These assigned spaces assume an essential part in saving biodiversity, giving territories to imperiled species, and protecting biological systems. The development and successful administration of safeguarded regions stay indispensable to worldwide preservation techniques.
2. **Species Preservation and Rebuilding:**
   Protection endeavors stretch out to the safeguarding of individual species confronting risk or annihilation. Species-centered drives incorporate hostage reproducing programs, environment rebuilding, and renewed introduction into nature. Examples of overcoming adversity, for example, the recuperation of the bald eagle and the dark footed ferret, highlight the effect of designated preservation measures.
3. **Global Joint effort and Arrangements:**
   Tending to worldwide ecological difficulties requires global joint effort. Different arrangements and shows, like the Show on Natural Variety (CBD) and the Assembled Countries Structure Show on Environmental Change (UNFCCC), give systems to facilitated activity. These arrangements stress the inter-

connectedness of biodiversity preservation, environmental change moderation, and manageable turn of events.

4. **Local area Based Preservation:**

Perceiving the significance of nearby networks in preservation, local area based drives enable occupants to play a functioning job in saving their regular environmental elements. These endeavors include practical asset the board, ecological training, and associations that adjust preservation objectives to neighborhood requirements and needs.

II. Effective Preservation Drives:

1. **Gorongosa Public Park Rebuilding, Mozambique:**
   The Gorongosa Public Park in Mozambique fills in as a remarkable illustration of fruitful protection and rebuilding. Following long stretches of contention, the recreation area went through a complete recovery exertion. The reclamation project zeroed in on environmental rejuvenation, local area commitment, and natural life security. Today, Gorongosa embodies the positive results attainable through incorporated preservation draws near.
2. **The Yellowstone Wolf Renewed introduction, USA:**
   The renewed introduction of wolves to Yellowstone Public Park in the US remains as a milestone progress in animal categories preservation. The renewed introduction in 1995 had flowing impacts on the recreation area's environment, prompting the recuperation of vegetation, changes in creature conduct, and expanded biodiversity. This drive displayed the complex connections inside environments and the potential for cornerstone species to drive positive biological movements.
3. **The Continuous Recuperation of the Humpback Whale:**

Protection endeavors zeroed in on the humpback whale have brought about prominent triumphs. Severe security measures, like the prohibition on business whaling, have added to the recuperation of humpback whale populaces. Cooperative examination, protection approaches, and public mindfulness crusades highlight the potential for purposeful endeavors to resuscitate undermined marine species.

III. Challenges in Worldwide Preservation Endeavors:

1. **Living space Misfortune and Fracture:**
   Living space misfortune and fracture stay imposing difficulties in protection. Human exercises, including deforestation, urbanization, and rural development, proceed to modify and reduce normal natural surroundings. Divided scenes present dangers to species that require enormous, interconnected territories for endurance.

2. **Environmental Change Effects:**
   Environmental change fuels existing preservation challenges by modifying biological systems, disturbing relocation designs, and expanding the recurrence and power of outrageous climate occasions. Species are compelled to adjust or relocate to new territories, putting extra weight on currently weak populaces.
3. **Poaching and Unlawful Untamed life Exchange:**
   Poaching and the unlawful untamed life exchange continue as significant dangers to numerous species, pushing some to the edge of termination. Notorious species like rhinos and elephants face the persevering quest for their important body parts. Powerful preservation techniques should incorporate strong enemy of poaching measures, policing, worldwide participation.
4. **Intrusive Species and Infections:**

Intrusive species and infections presented by human exercises present critical dangers to local biological systems. These trespassers can outcompete nearby species, upset food networks, and add to decreases in biodiversity. Preservation endeavors should address the counteraction, control, and the board of intrusive species to safeguard local verdure.

IV. Future Headings in Preservation:

1. **Bridling Innovation for Observing and Authorization:**
   Headways in innovation, including satellite symbolism, drones, and computerized reasoning, offer remarkable open doors for observing and authorizing preservation endeavors. Remote detecting advancements empower constant following of living space changes, criminal operations, and untamed life populaces, upgrading the proficiency and adequacy of protection drives.
2. **Genomic Preservation and Hereditary Variety:**
   Genomic preservation, including procedures, for example, hereditary banking and cloning, holds guarantee for saving hereditary variety and forestalling the deficiency of remarkable attributes inside species. Hereditary mediations might assume a part in the recuperation of imperiled populaces, improving their versatility to natural changes.
3. **Rebuilding Environment and Rewilding:**
   Rebuilding environment and rewilding methodologies center around restoring corrupted biological systems and once again introducing local species to their authentic reaches. These methodologies mean to turn around the effects of environment misfortune and discontinuity, cultivating self-supporting biological systems that can uphold assorted verdure.
4. **Preservation Money and Feasible Turn of events:**

Getting subsidizing for protection drives stays vital. Preservation finance models, including ecotourism, installment for biological system administrations, and effect effective financial planning, offer creative ways of producing income for protection while supporting manageable turn of events. Adjusting protection objectives to monetary motivators makes a mutually beneficial situation for biodiversity and nearby networks.

V. Connecting with Neighborhood People group and Native Information:

1. **Engaging Nearby Stewardship:**
   Connecting with nearby networks in protection endeavors cultivates a feeling of pride and stewardship. Engaging occupants to effectively take part in navigation, practical asset the board, and protection programs guarantees that drives line up with nearby necessities and needs.
2. **Incorporating Native Information:**

Integrating Native information and viewpoints advances preservation systems. Native people group frequently have conventional environmental information that is priceless for figuring out biological systems, distinguishing key species, and creating supportable practices. Cooperative associations with Native gatherings advance social variety and upgrade preservation results.

VI. Instruction and Support for Practical Living:

1. **Natural Instruction Projects:**
   Natural instruction stays an intense device for cultivating mindfulness, understanding, and obligation to preservation. Carrying out instructive projects at different levels, from schools to local area studios, develops a protection mentality and furnishes people with the information to make educated, feasible decisions.
2. **Promotion for Strategy Change:**

Backing assumes a urgent part in affecting strategy choices that influence preservation. Participating out in the open talk, utilizing web based entertainment, and supporting backing associations add to bringing issues to light and earning public and political help for preservation arrangements. Resident commitment is a strong power for driving positive change.

VII. The Job of Organizations and Enterprises in Protection:

1. **Corporate Social Obligation (CSR) Drives:**
   Organizations and partnerships can possibly be strong partners in protection endeavors. Coordinating preservation into corporate social obligation drives can prompt interests in reasonable practices, territory security, and the help of

protection associations. Exhibiting a promise to natural stewardship improves corporate notorieties and adds to positive protection results.

2. **Feasible Stockpile Chains:**

Taking on reasonable store network rehearses is fundamental for ventures that depend on normal assets. Executing mindful obtaining, lessening natural effect, and advancing straightforwardness all through supply anchors add to protection objectives by limiting territory annihilation and environmental corruption.

VIII. Moral Contemplations in Protection:

1. **Local area and Partner Inclusivity:**
   Moral protection rehearses focus on inclusivity and joint effort with nearby networks and partners. Drawing in networks in dynamic cycles, regarding social qualities, and recognizing customary privileges add to moral and viable protection drives.
2. **Creature Government assistance in Protection Mediations:**
   Protection intercessions, for example, hostage rearing and renewed introduction programs, should focus on the government assistance of individual creatures. Moral contemplations incorporate guaranteeing proper everyday environments, limiting pressure, and focusing on the wellbeing and prosperity of creatures associated with protection drives.
3. **Adjusting Human Necessities and Protection Objectives:**

Finding some kind of harmony between human necessities and protection objectives is a complex moral test. Moral protection rehearses consider the financial setting of neighborhood networks, planning to ease neediness, further develop vocations, and address human-natural life clashes in manners that help both human prosperity and biodiversity preservation.

IX. Difficulties and Open doors in Future Protection Endeavors:

1. **Worldwide Joint effort and Political Will:**
   Conquering worldwide protection challenges requires supported joint effort and political will. Global participation is fundamental for resolving issues that rise above public boundaries, for example, environmental change, unlawful natural life exchange, and territory obliteration.
   Empowering political pioneers to focus on preservation and allot assets is vital for powerful worldwide activity.
2. **Tending to Imbalance and Differences:**
   Preservation endeavors should address financial imbalances that frequently add to natural corruption. Impartial circulation of the advantages of preservation,

fair asset the board, and neediness lightening are fundamental parts of making supportable and just protection results.

3. **Adjusting to Arising Dangers:**

As the world advances, new dangers to biodiversity might arise. Fast urbanization, mechanical progressions, and unexpected environmental disturbances present difficulties to customary protection draws near. Future endeavors should stay versatile and receptive to arising dangers, consolidating state of the art research and inventive arrangements.

9.1Overview of kangaroo conservation status

Kangaroos, famous marsupials local to Australia, are basic to the mainland's one of a kind biological systems. The preservation status of kangaroos is a subject of huge worry as human exercises, environmental change, and living space misfortune keep on influencing their populaces. This exhaustive outline investigates the ongoing preservation status of kangaroos, the difficulties they face, fruitful protection drives, and the future points of view for guaranteeing the endurance and prosperity of these entrancing animals.

1. **Kangaroo Species and Their Dispersion:**

1. **Red Kangaroo (Macropus rufus):**
   The red kangaroo is the biggest marsupial and is far and wide across the greater part of Australia. Known for its unmistakable ruddy earthy colored fur, this species possesses different scenes, from prairies to deserts.
2. **Eastern Dark Kangaroo (Macropus giganteus):**
   The eastern dark kangaroo is tracked down in the eastern and southern pieces of Australia, leaning toward woods and meadows. With a dark earthy colored coat and unmistakable facial markings, this species is versatile to a scope of conditions.
3. **Western Dim Kangaroo (Macropus fuliginosus):**
   The western dim kangaroo is conveyed across the southwestern pieces of Australia. Comparative in appearance toward the eastern dim kangaroo, it flourishes in waterfront locales, forests, and shrublands.
4. **Antilopine Kangaroo (Macropus antilopinus):**
   The antilopine kangaroo is local to northern Australia, principally occupying savannas and prairies. With a ruddy earthy colored coat and unmistakable markings, this species shows one of a kind ways of behaving contrasted with its southern partners.
5. **Different Species:**

Furthermore, there are more modest kangaroo species like the wallaroos (Macropus robustus) and tree kangaroos (Dendrolagus spp.), each adjusted to explicit living spaces and showing different environmental jobs.

II. Protection Status of Kangaroos:

1. **IUCN Red Rundown Appraisals:**
   The Worldwide Association for Preservation of Nature (IUCN) surveys the protection status of species universally. While individual kangaroo species might have fluctuating orders, the general pattern recommends that most kangaroo species are not presently delegated jeopardized. Notwithstanding, explicit populaces and subspecies might confront limited dangers.
2. **Dangers to Kangaroo Populaces:**

Regardless of their overall characterization as types of least concern, kangaroo populaces face different dangers that could affect their drawn out endurance:

Territory Misfortune and Fracture: Urbanization, horticulture, and framework advancement add to natural surroundings misfortune and discontinuity, segregating kangaroo populaces and restricting their admittance to assets.

Environmental Change: Adjusted weather conditions, outrageous temperatures, and changes in vegetation because of environmental change can influence the accessibility of food and water, affecting kangaroo natural surroundings.

Human-Untamed life Struggle: Kangaroos frequently clash with human exercises, particularly in regions where horticulture and urbanization cross with their regular living spaces. This contention can prompt negative discernments and, at times, winnowing programs.

Sickness and Parasites: Untamed life illnesses and parasitic contaminations can present dangers to kangaroo populaces, particularly in regions where human and kangaroo living spaces cross-over.

III. Effective Protection Drives:

1. **Local area Drove Protection Projects:**
   Local area drove drives assume an essential part in kangaroo protection. Neighborhood people group, alongside preservation associations, frequently carry out programs that attention on living space rebuilding, untamed life passageways, and government funded schooling to cultivate concurrence.
2. **Salvage and Restoration Focuses:**
   Salvage and restoration focuses contribute fundamentally to the protection of kangaroo species. These offices salvage harmed or stranded kangaroos, give clinical consideration, and, whenever the situation allows, discharge them back into nature. This approach mitigates the effect of human-natural life struggle and supports individual kangaroos in trouble.

3. **Preservation through Native Land The board:**
   Native people group have a significant association with the land and frequently take part in preservation rehearses that line up with customary biological information. Cooperative endeavors between Native people group and protection associations advance maintainable land the executives, helping both kangaroo populaces and biological systems.
4. **Examination and Observing Projects:**

Progressing examination and observing projects contribute significant information to figure out kangaroo conduct, populace elements, and biological prerequisites. This information illuminates preservation procedures, assisting with tending to explicit dangers and carry out designated intercessions.

IV. Challenges in Kangaroo Protection:

1. **The executives of Human-Untamed life Struggle:**
   Human-untamed life struggle stays a critical test in kangaroo protection. In locales where kangaroos connect with horticultural exercises, there are clashes over brushing lands, prompting negative discernments and, in outrageous cases, government-authorized winnowing programs.
2. **Supportable Land Use Practices:**
   Adjusting the necessities of human advancement with the preservation of kangaroo living spaces requires maintainable land use rehearses. As urbanization and agribusiness grow, finding arrangements that oblige both human and untamed life needs turns out to be progressively complicated.
3. **Environmental Change Effects:**
   Environmental change represents a complex danger to kangaroo populaces. Modified atmospheric conditions, delayed dry seasons, and changing vegetation can influence the accessibility of food and water assets. Variation techniques should be carried out to alleviate these effects.
4. **Worldwide Interest for Kangaroo Items:**

The business utilization of kangaroo items, including meat and calfskin, raises moral and preservation concerns. While managed gathering is legitimate in certain districts, the global interest for these items might possibly prompt overexploitation and adverse results for kangaroo populaces.

V. Future Bearings in Kangaroo Preservation:

1. **Incorporated Scene The executives:**
   Future preservation endeavors ought to focus on coordinated scene the board, taking into account the interconnectedness of biological systems. This approach includes making untamed life hallways, safeguarding basic natural surroundings,

and advancing reasonable land use rehearses that benefit both kangaroo populaces and the more extensive climate.

2. **Local area Commitment and Instruction:**
   Local area commitment and instruction drives are significant for cultivating understanding and backing for kangaroo preservation. Public mindfulness crusades, instructive projects, and local area contribution in protection navigation add to a more educated and steady society.
3. **Strategy Support for Feasible Practices:**
   Upholding for strategies that advance reasonable practices is fundamental. Preservation associations and activists can pursue impacting government strategies connected with land use, untamed life the executives, and human-untamed life compromise to guarantee a fair and feasible way to deal with kangaroo protection.
4. **Research on Environmental Change Flexibility:**

Given the effects of environmental change on kangaroo natural surroundings, research zeroed in on environmental change flexibility is significant. Understanding how kangaroo populaces adjust to changing circumstances can illuminate preservation techniques that improve their capacity to flourish in a powerful climate.

VI. Moral Contemplations in Kangaroo Preservation:

1. **Creature Government assistance in Preservation Activities:**
   Moral contemplations in kangaroo preservation include focusing on the government assistance of individual creatures. This remembers guaranteeing sympathetic treatment for salvage and recovery programs, tending to stressors in hostage conditions, and limiting the adverse consequences of protection mediations on kangaroo prosperity.
2. **Native Privileges and Social Contemplations:**
   Regarding Native privileges and social contemplations is major to moral kangaroo protection. Cooperative methodologies that recognize and integrate Native information, values, and practices add to the general achievement and manageability of protection drives.
3. **Offsetting Protection with Financial Interests:**

Moral kangaroo preservation likewise requires finding some kind of harmony between protection objectives and financial interests. This includes considering the financial requirements of nearby networks and carrying out protection procedures that give benefits without compromising the prosperity of kangaroo populaces.

9.2Role of captive populations in conservation

The preservation scene has developed, requiring creative ways to deal with defend jeopardized species and keep up with biodiversity. Hostage populaces, housed in zoos, aquariums, and rearing offices, assume an essential part in this change in outlook. Past giving diversion to guests, these hostage populaces act as instrumental devices in protection endeavors, adding to research, training, and the possible rebuilding of undermined species in their normal territories.

1. **Hereditary Variety and Species Protection:**
   Hostage populaces act as hereditary repositories, saving the variety of species confronting up and coming dangers in nature. In circumstances where wild populaces are lessening because of environment misfortune, environmental change, or poaching, hostage reproducing programs become crucial. Keeping a different hereditary pool inside these projects guarantees that significant hereditary characteristics are not lost, giving a source to expected renewed introduction or support of wild populaces.
2. **Species Recuperation and Renewed introduction Projects:**
   Hostage reproducing programs go about as a fundamental part of species recuperation and renewed introduction drives. At the point when an animal categories faces the edge of termination, laying out a steady hostage populace turns into a help.
   Through painstakingly overseen reproducing and farming practices, hostage populaces can deliver people that are subsequently once again introduced into their normal natural surroundings. This cycle supports reinforcing existing populaces, reestablishing biological systems, and alleviating the effect of human exercises.
3. **Research Open doors and Preservation Bits of knowledge:**
   Hostage populaces offer a controlled climate for specialists to lead top to bottom investigations on species conduct, physiology, and generation. Experiences acquired from hostage studies give significant information that can illuminate protection techniques. Figuring out the wholesome necessities, conceptive science, and social elements of an animal types in imprisonment adds to better administration rehearses both inside hostage offices and in nature.
4. **Training and Public Mindfulness:**
   Zoos and aquariums, lodging hostage populaces, act as strong instructive stages. By exhibiting jeopardized species very close, these offices encourage an association among guests and untamed life, imparting a feeling of obligation for preservation. Instructive projects and interpretive displays upgrade public mindfulness about the difficulties confronting species in the wild, reassuring a guarantee to protection endeavors.
5. **Crisis Reaction to Basic Circumstances:**
   Hostage populaces go about as a security net in crisis circumstances, like illness flare-ups, cataclysmic events, or unexpected decreases in wild populaces. In occasions where prompt activity is expected to forestall termination, people from

hostage populaces can be sent for support or renewed introduction, guaranteeing the endurance of the species.

6. **Versatile Administration and Exploration Advancement:**
   Hostage populaces give a controlled climate to testing and refining preservation methodologies. Specialists can explore different avenues regarding the board methods, dietary mediations, and conceptive advances in bondage, applying fruitful ways to deal with in-situ preservation programs. This versatile administration approach considers advancement and refinement before execution in the frequently flighty states of nature.
7. **Worldwide Coordinated effort and Information Trade:**

Hostage populaces work with worldwide coordinated effort among zoos, research establishments, and protection associations. Information trade, shared reproducing programs, and cooperative examination endeavors upgrade the aggregate comprehension of species science and the board. This interconnected organization reinforces protection drives by pooling assets, aptitude, and encounters.

Challenges and Moral Contemplations:

While the job of hostage populaces in protection is significant, it accompanies difficulties and moral contemplations. Keeping up with the physical and mental prosperity of hostage creatures is foremost. Satisfactory nook sizes, advancement exercises, and social communications are fundamental for advancing normal ways of behaving and guaranteeing the general government assistance of people. Also, the choice to once again introduce hostage reproduced people should be founded on careful logical appraisals, taking into account environmental variables and the availability of the populace to flourish in nature.

9.3Future challenges and opportunities

As we peer into the fate of preservation, a scene set apart by the two difficulties and potential open doors unfurls. The multifaceted snare of natural, social, and innovative elements presents a perplexing embroidery that progressives should explore. This investigation digs into the expected difficulties and valuable open doors that lie ahead, offering bits of knowledge into how the field of protection can adjust and improve to shield biodiversity and environmental honesty.

****1. Environmental Change and Territory Adjustments:**

Challenges:

Environmental change represents a significant danger to biodiversity, modifying biological systems and influencing the dispersion and conduct of species. Climbing temperatures, outrageous climate occasions, and moving precipitation examples can prompt natural surroundings misfortune, disturbing the fragile equilibrium that supports numerous species. As natural surroundings change, species might confront difficulties in adjusting or relocating to additional reasonable conditions.

Open doors:

Relieving and adjusting to environmental change is a chance for preservation development. Embracing feasible practices, like reforestation, carbon sequestration, and the advancement of strong biological systems, can upgrade the capacity of natural surroundings to endure environment related pressure.

Protectionists can team up with networks to execute environment savvy methodologies that advance both biodiversity and human prosperity.

****2. Human-Untamed life Struggle:**

Challenges:

The infringement of human exercises into normal living spaces frequently prompts expanded collaborations among people and natural life. This can bring about clashes as species, especially enormous well evolved creatures, look for assets in closeness to human settlements. Such contentions might prompt negative impression of natural life, bringing about retaliatory killings, living space corruption, and further discontinuity.

Potential open doors:

Tending to human-untamed life struggle presents a chance for inventive preservation arrangements. Executing people group based preservation drives that integrate neighborhood points of view and needs can encourage concurrence. The improvement of advances, like early admonition frameworks and non-deadly hindrances, can assist with forestalling clashes and safeguard both human occupations and natural life.

****3. Loss of Biodiversity and Environment Administrations:**

Challenges:

The speeding up loss of biodiversity sabotages the complicated trap of cooperations that support biological systems. As species vanish, fundamental natural capabilities might be compromised, influencing environment administrations indispensable for human endurance. Decreases in pollinators, water purifiers, and supplement cyclers can upset the sensitive equilibrium that upholds agribusiness, water supply, and generally speaking biological system wellbeing.

Open doors:

Preservation endeavors can gain by chances to reestablish and restore environments. Reforestation projects, wetland reclamation, and feasible land the board rehearses add to the recuperation of biodiversity and the arrangement of environment administrations. Perceiving the worth of biodiversity in supporting human prosperity and monetary exercises cultivates a common obligation regarding protection.

****4. Mechanical Progressions:**

Challenges:

While innovative progressions offer uncommon devices for preservation, their execution accompanies difficulties. The abuse of innovation, like the unlawful utilization of robots, hereditary designing, or man-made reasoning, can present dangers to untamed life. Finding some kind of harmony between mechanical development and moral, dependable use is pivotal for protection achievement.

Potential open doors:

Innovation gives a variety of chances to upgrade preservation endeavors. Remote detecting advancements, satellite symbolism, and robots empower ongoing observing of biological systems, helping with territory the executives and insurance. Hereditary advances offer opportunities for protecting hereditary variety and tending to sickness dangers. Embracing development while complying with moral rules guarantees that innovation turns into a strong partner in protection.

****5. Worldwide Availability and Joint effort:**

Challenges:

Protection challenges frequently rise above public lines, requiring worldwide co-operation. In any case, political and financial interests might thwart powerful participation. Abberations in assets, varying protection needs, and international pressures can obstruct aggregate endeavors to resolve worldwide natural issues.

Amazing open doors:

Worldwide network gives an open door to the trading of information, assets, and cooperative endeavors. Peaceful accords, like the Show on Organic Variety (CBD) and the Paris Understanding, act as structures for composed activity. Arising drives, as transboundary protection tasks and joint exploration attempts, exhibit the potential for shared liability and aggregate arrangements.

****6. Human Populace Development and Urbanization:**

Challenges:

The growing human populace and quick urbanization put expanding tension on normal environments. Endless suburbia, foundation improvement, and the interest for assets add to territory fracture and misfortune. As human populaces develop, so does the requirement for land, water, and energy, escalating the opposition between human necessities and natural life preservation.

Potential open doors:

Practical metropolitan preparation and improvement offer chances to offset human requirements with preservation objectives. Executing green foundation, making untamed life hallways inside metropolitan scenes, and coordinating biodiversity contemplations into city arranging add to conjunction. Training and mindfulness missions can cultivate a protection ethic among metropolitan populaces, advancing dependable connections with the regular world.

****7. Moral Contemplations and Protection Equity:**

Challenges:

Moral contemplations in protection include exploring complex inquiries connected with creature government assistance, Native privileges, and the fair circulation of preservation benefits. Adjusting the necessities of nearby networks with preservation objectives, tending to social contemplations, and guaranteeing the moral treatment of individual creatures in hostage populaces present continuous difficulties.

Open doors:

Incorporating morals into preservation rehearses presents an amazing chance to upgrade the decency and viability of protection drives. Cooperative associations with Native people group, conscious commitment with nearby information, and focusing on the government assistance of the two people and natural life add to preservation equity. Moral contemplations guarantee that preservation endeavors line up with standards of social and ecological obligation.

www.ingramcontent.com/pod-product-compliance
Lightning Source LLC
LaVergne TN
LVHW010432230826
846092LV00009BA/1135

9788196832247